Peter Marsh is Industry Editor of N
special responsibility for computing
investigates the revolution that microprocessors have brought about in society, and the implications that these changes have for civilization in this century and the next.

Peter Marsh

THE SILICON CHIP BOOK

ABACUS

First published in Great Britain in ABACUS, 1981
by Sphere Books Ltd
30/32 Gray's Inn Road, London, WC1X 8JL

Set in Lasercomp Times

Printed and bound in Great Britain by
©ollins, Glasgow

Contents

Tables

Figures

Notes

All cash figures in 1979 dollars or pounds except where stated otherwise, or where text refers to a particular year in which case figures will be at current prices. Conversion rate in 1979 taken to be £1 = $2.2. One billion is 1000 million.

Part 1

TECHNOLOGY

Chapter 1: What is microelectronics?

The Moon landing of 20 July 1969 was a remarkable moment in the history of mankind. For the first time, man stepped foot on territory other than the Earth. The story of that exploit by the American astronauts Neil Armstrong and Edwin Aldrin has been told many times already and will not be dwelt upon here. This book is about something seemingly more prosaic: the advent of small, cheap electronic components. But these tiny bits of hardware will change our lives. And the forces that engendered them were very much concerned with that adventure of the Moon landing. It was tiny electric components – lots of them, manufactured cheaply, cleverly and to very high standards of reliability – that made the Moon mission possible. During the 1960s, America's burgeoning space industry spent millions of dollars on the electronic parts that would control the spacecraft on its 400,000 km journey and bring its occupants safely home.

Similar electronic parts are now making their mark on Earth. When people in the 1960s asked of the 'space race', 'Yes, it's marvellous, but how will it affect us?', the answer might have been: 'It will bring the world cheap computers.'

Small, cheap electronic components are important mainly because they provide the means to produce small, cheap computers. The technique of 'shrinking' electronic components to the degree possible today is called microelectronics. It can be looked upon as a form of miniaturisation. This is a phenomenon that affects many areas of life more or less unheralded, certainly without having a book written about it. After all, smallness has many obvious advantages; it confers cheapness, portability and the ability to squeeze things into a confined space. To take a simple example, miniaturisation provides beer cans so people can have a drink at home without the bother of having to heave around a barrel. But in no other sphere of human activity has size reduction been as important as in electronics. For one thing, the most useful of all the devices that can be made from electronic parts – computers – have a greater potential to alter life than any other machine invented. (A few years ago the internal combustion engine and the electric motor might have rivalled the computer for this accolade, but now they have been well beaten into second and third places.)

Second, the miniaturisation that has affected electronic components, and with them computers, has been on a scale seldom seen before. A simple example shows the enormity of the change. When the ENIAC machine, the world's first useful electronic computer, appeared after World War II it weighed thirty tonnes, took up the floor space of a church hall, and consumed the power of a steam train. A computer of similar power today fits into a few pieces of semiconductor material each the size of a shirt button (the so-called silicon chip). And with this reduction in dimensions has come a 'shrinking' in price of staggering proportions. Today's machine costs perhaps £100 compared with the £1.5 million of the hulking monster of the 1940s.

Such a story could almost have come from a fairy tale. If similar events occurred in books obviously centred on fantasy, such as *Gulliver's Travels* or *Alice in Wonderland*, the readers could be excused for thinking the authors had stretched credibility a little too far. So how has such an epic tale of miniaturisation occurred amid the harshness of real life?

To start to answer this, we must introduce an entity of great significance for the whole of this book. This is the electron, one of the basic building blocks of matter. The electron may be considered a ball of tiny dimensions that measures roughly one five-million-millionths of a centimetre in diameter. The world that the electron inhabits is almost unimaginable because it is so small. But what happens in this world is crucial. The movement of electrons from one point to another produces the phenomenon of electricity. This drives motors, heats water, and provides light in almost every part of existence. Electronics, or the science of electrons, is the study of what happens to electrons under different conditions. Electronic engineers are the people who try to understand and control the forces that influence these tiny particles. They will figure greatly in this book: electrons are to these people what stars are to astronomers.

The reader may well ask: 'But the world of electrons is small enough already; surely it cannot be reduced any more?' The answer is that, in microelectronics, electrons still perceive the same world as they have always done. What are shrunk, however, are the devices, or components, that electronic engineers invent to contain the electrons and make their activities useful. To return to the analogy of the beer can and the barrel, the beer has the same characteristics in either receptacle. But in the smaller one it is often in a much handier form, as anyone who wants a quick drink, but doesn't have the fork-lift truck needed to bring the barrel to his lips, will tell you.

Something similar applies to electrons. They are generally more useful in a 'shrunk', or microelectronic device, than in a conventionally-sized electronic component. Well-known examples of ordinary electronic devices are vacuum valves, transistors, resistors, capacitors and amplifiers. Look inside a 1950s TV or radio set and you will see lots of them. Such devices are very useful things. The existence for many years of all the common pieces of electrical equipment that clutter up our homes and workplaces testifies to this. But what if they could somehow be made hundreds, if not thousands, of times smaller and cheaper, while still performing the same functions? The usefulness of these components would be magnified enormously as a result. Something to this effect has indeed happened with the production of microelectronic components, although they are rather more difficult to make than by immersing the old-style electronic devices in some sort of 'shrinking medium, as we shall see later.

Computers are made from the same type of electronic components as you will find tucked away in TVs and radios and so on. There is one important difference, however. Whereas the most important bits of computers rely almost exclusively on lots of electronic components to make them work, such things are only a relatively small part of a TV set, or a radio, or an electric cooker. Easily the biggest part in these devices are objects like TV tubes, aerials, tuning devices, loudspeakers, hot-plates, and baking pans. And these are devices that cannot be magically 'shrunk' by microelectronics. This is one of the reasons that this book is concerned chiefly with the miniaturisation of computers, not TVs or electric cookers or, for that matter, washing machines or hairdryers.

The second reason is simply that computers are more important than these other more humble pieces of electrical machinery. This is bound to raise the hackles of people who make washing machines, hairdryers or electrical sewing machines. But the fact can hardly be disputed. There is nothing particularly clever about computers: rather, their great attributes, as we shall explain later, are speed and versatility. Where the cleverness comes in is in humans dreaming up ways of capitalising on computers' versatility. People have shown magnificent resourcefulness in thinking of uses for computers. When the last count was made, there were some 25,000; and, no doubt, by now there are a lot more.

This is where the technology of microelectronics becomes so important. While computers were large, expensive, unreliable machines constructed out of orthodox electronic components, most of these uses were bound to go unrealised. It was only as the size and cost of computers started to drop, brought about by gradual advances in microelectronics, that these uses started to become apparent. Microelectronics has transformed computers to the point that they have become the everyday objects they always promised they would be.

Just why are computers so important? A computer can best

be thought of as a 'decision-making machine'. This is not to say it makes decisions for itself. Rather it makes decisions for its operators, who are human beings. An analogy with people helps here. Think of a particularly inept person, who has a certain amount of intelligence (thinking power) but cannot make up his mind about anything. He can never decide where to go in the evening, where he wants to go on holiday, which route to take to work or which clothes to put on in the morning. This person's life would be a complete disaster had not a friend worked out a method by which he could have decisions made for him.

This method would take into account all the variables that affect the man's life. It would rate them according to a set numerical scale, and process them in a way that all the factors were carefully weighed and the most appropriate decision made. For instance, to decide where to go for a night's entertainment, the sorry individual would have to jot down 'scores' for variables such as the films and other events on in his locality, the various admission prices, weather conditions, what sort of mood he is in, whether any of his friends are interested in going with him and so on. The 'method' would be a series of mathematical equations which would account for all the variables and feed the man with a decision, in much the same way as his brain would do were he capable of using it.

The analogy, though far from exact, helps us understand how a computer works. A computer is a machine that takes in different numbers coded so that they mean something to the outside world (they mean nothing to the computer), treats them mathematically and then churns out the result. A computer would be a very useful instrument for the person just described. He could arrange for it to handle all the variables governing his life to tell him what was most appropriate to do in any given set of circumstances.

The great thing about a computer is that the way it handles numbers can be changed fairly easily. This is called programming a computer. A change in programming lets the computer operator alter the role of his machine. Thus the hopeless, dithering person we have just mentioned would have to change his computer's 'program' (as the machine's set of operating instructions is called) to order the machine to do different tasks. For instance, the program instructing the computer to map out his route to work in the morning would be different from the one telling him how to come home at night.

Of course, computers can carry out far more complex tasks than these. Mostly they are useful for arriving at very complex decisions very quickly. Thus computers were crucial in the 1969 Moon mission because they made decisions on a wealth of complicated data (such as speed and orientation of the spacecraft, functioning of the rocket motors, the state of the astronauts' health, the position of stray meteorites in the spacecraft's path) much more quickly and adroitly than a person could ever do. Some idea of the applications for computers may now perhaps be appreciated. Wherever there are variables to be handled and processed, or decisions to be made, a computer can be slotted into place.

The most obvious applications are for straightforward mathematical calculations, for instance working out (making decisions on) complex equations in science. But computers also have a place in the home, for working out tax bills, totting up what shopping should be done during the week, or regulating heating to make the house comfortable but maximise energy saving. They have a place in processing industry, to make decisions on when to open and shut valves for instance, or how much of ingredient X to add to ingredient Y at an appropriate time. Computers have a huge potential in manufacturing industry, where they could take account of many variables to decide how often to add a bolt in a particular place to a component on an assembly line. They will be seen increasingly inside cars, to regulate for instance carburettors. They are being applied more and more in offices (inside computerised typewriters called word processors which 'decide' on things like text editing and whether a typist has spelt a word incorrectly); and in telecommunications where the machines make thousands of decisions every second to speed computerised messages to their correct destinations. Many of the applications will be looked at in detail later in this book.

How can a machine have such powers? We must remember that a computer is nothing more than a processor of numbers. Before it can work on a decision-making process, the computer must transform it into a set of arithmetic steps such as addition, multiplication and subtraction. Computers have a special logic code which enables them to change ideas from the outside world into numbers – we shall have more to say on how they do this in Chapter 2. Groups of electronic circuits called logic circuits carry out the arithmetic steps to arrive at the computer's 'decision'.

We can look to ordinary life to find a simple mechanism of numbers representing ideas – the Morse code. Two people who know the code can communicate with each other by tapping out a set number of 'dots' and 'dashes' over a radio link, or by flashing the sequence with a torch as a series of bursts of light. The Morse code is a particularly useful metaphor because, like the Morse operator, the computer has for its basic 'language' a code consisting of just two 'statements' similar to the dot and the dash. The statements that a computer recognises are whether an electric current – a stream of electrons – is either on or off. In computer terms, current 'on' corresponds to the number 1, and 'off' to the number 0.

We can now see how simple the basic operating process of a computer is. Each arithmetic-processing part of a computer can be likened to a stupid animal that, of all the events happening in the world around it, can only recognise two things, night and day. Because they operate on a code in which only two statements are important, the computer, the Morse operator and the thick-witted animal can all be said to recognise a 'binary' code (binary comes from the Latin word *bini* meaning 'two together').

To human eyes, the binary code for computers looks pretty

meaningless. It consists of a stream of 0s and 1s transmitted as electronic pulses that are so arranged that they mean something to the computer. A computer is very quick at understanding its admittedly limited vocabulary of 'bits' (as the 0s and 1s, or binary digits, are called). It has no trouble in making sense of a bit stream transmitted at 10,000 digits-per-second, or 1000 instructions in a blink of an eyelid. The constitutent arithmetic-processing parts of the fastest computers understand their diet of instructions so quickly that they can absorb 10^{10} binary digits per second. This speed is almost beyond understanding. Let us imagine a person sets out to assimilate this number of instructions at a rate of six a minute, or one every ten seconds, a reasonably fast rate you might think. Then if he worked flat-out twelve hours a day, seven days a week, the job would take him 6000 years. If the person started the job about the time Stonehenge was being planned, he would still have another 1000 years of work before the task was finished. Now, perhaps, we can realise why it is that computers are so useful, even if the way in which they work is fairly simple. It is that their work rate is so phenomenally high.

Electronic computers have had a very short history. They didn't appear in practice until after World War II, though people thought of the ideas that led to them centuries before this. The arithmetic processing in the early machines' logic circuits was done by thousands of big electronic components (valves, resistors, capacitors and so on) linked by miles of tangled wires. Because of their complex construction, the early machines (such as the ENIAC) were unreliable and broke down easily. The first advance came in 1947 when physicists in America showed how the properties of the valve could be simulated by a tiny piece of material called a transistor. The transistor set electronic engineers on the road to microelectronics because it introduced the idea that semiconductor materials (of which transistors are made) could be useful in designing electronic components. The best known, and most valuable, semiconductor is silicon, the element in sand. Semiconductors have unusual properties that make it simple to control the flow of electrons through them by changing a voltage level. With a series of transistors, engineers can construct the logic circuits needed in computers, which before the advent of semiconductor technology had to be made from valves. The important difference between a transistor and a valve is that the semiconductor device is ten to twenty times smaller, consumes far less power and is less likely to go wrong.

The next important step was in the late 1950s when engineers found they could join components such as the transistor, the resistor and the capacitor on one piece of semiconductor, a process that was called 'integration'. The integrated circuit was born, that is, a circuit with all its components on one tiny substrate, or chip, of silicon. With the help of new technology, electronic firms started to produce pieces of semiconductor containing the equivalent of perhaps four or five components. This was pretty good going, of course, compared with the one-component-on-a-chip transistor, and little short of marvellous compared with the days of the valve. But after this electronic engineers, a very young breed of people who had hardly existed before 1940, really started to show the world how clever they could be. They managed to squeeze more and more components on to the same small chip until we have the situation today, when you can find chips half a centimetre square containing as many as 200,000 electronic components, and costing a few pounds. It was a series of incredible achievements, which will be explained in more detail later in the book.

Of today's chips, the most important is the microprocessor. It contains all the logic circuits that make up the arithmetic-processing 'brains' of a modern computer. Other chips are needed to make a fully-functioning computer. For instance, the memory chip contains data, stored as a series of binary pulses, that the microprocessor chip acts upon. The memory provides the latter with information in a similar way as a mass of writing in a book presents information to a person's brain when he reads it. Other devices, called input and output chips, arrange for chunks of binary code to be pulled out of, or inserted in, the processor chip at appropriate times. It is now possible to make a chip that is so densely packed with circuitry that it contains the functions of processor, memory, and all the other bits and pieces of a computer. This is the single-chip microcomputer: the term 'microcomputer' without such a prefix is given to a small computer made up of a number of chips.

The magnitude of the reduction in size and cost of computers since they were first introduced takes a little while to sink in. Let us try to ram the message home by comparing one of the early commercial computers, the IBM 650 of the mid-1950s, with a 1978 Texas Instruments' programmable calculator of roughly equivalent power. The former fitted into seven cubic metres, the size of a large wardrobe, and weighed more than two tonnes while the calculator is a box 100,000 times smaller and 1000 times lighter. The IBM machine cost $200,000 in 1955 prices while the price of the calculator, expressed in 1955 dollars, is some 15,000 times less. Let us imagine the same 'scaling down' process applied to the vehicle that blasted America's astronauts to the Moon in 1969, the mighty Saturn V rocket that weighed 3000 tonnes and was as tall as a thirty-storey building. If this had happened, man would have gone to the Moon in a unit the size of a filing cabinet that weighed no more than a small car. And the cost of the rocket would have reduced from $400 million to some $30,000, equivalent to the price of a small house in England.

Cheap, everyday computers are new phenomena for the world to cope with. The world has grown used to devices that supplement muscle power: since cavemen made the first stone chisel, people have constantly found ways of improving on the physical powers of the human frame. The lever, Archimedes' screw, the plough, the water wheel, the stage coach, printing machines, steam engines, the spinning loom, the motor car, chemical plants, oil rigs, atomic reactors, nuclear bombs – all are devices which, in their various ways, save man the job of physically accomplishing something either difficult, impossible or merely tiresome for him to carry out on his own. But machines that make decisions like computers are different. Instead of adding to muscle power, they supplement (or supplant) *brain* power.

There have certainly been inventions before the computer that can be claimed to have helped people's thought processes. The printing press does this because it captures a lot of details in permanent form and saves people having to remember them. The telephone, TV, cinema and the radio are aids to thinking in the same kind of way. But there have not been many inventions which carry out 'thinking' on their own. The word has to be put in apostrophes because even the most advanced computers, have no way of knowing the significance of their activities, and do only what a human tells them to. But as people often make decisions by applying, albeit subconsciously, set rules of reason, and this process is called thinking, then we may say the same of computers, with only little trepidation that we are using the language wrongly. Thinking machines, becoming all the time cheaper, tinier and more unobtrusive to the point that before long we will probably forget that they are there altogether, offer a great many uses and problems for the world, and we shall discuss these in detail later in the book.

Chapter 2: The rise of the computer

History is peppered with episodes which show that people are good at finding uses for what technology makes possible, even though the uses are often scarcely evident before the technology appears. Take the invention, for instance, of the wheel, the printing press, the telephone and the TV. When these devices first appeared, you can be sure most people did not know what to do with them – but a few years later they could hardly get along without them. Nothing illustrates better this human tendency of failing to anticipate the future than the story of the computer.

The first electronic computers, the true predecessors of today's button-sized microprocessors, appeared just after World War II. The men who made them did so mainly because they thought the machines would cut the time taken to solve abstruse mathematical equations, the sort scientists used to think nothing of spending days struggling over. But few of the early computer builders put their machines' potential much beyond that. Arthur Porter, one of the real computer pioneers, who in the 1930s worked on a very early computer called the Manchester differential analyser, recalls now: 'I didn't feel there would be much scope for computers. I just didn't think they would be sufficiently reliable.' In the early days, computers certainly weren't. They used miles of wires, thousands of clumsy electrical valves, they consumed the power of a steam locomotive, and they broke down fearfully easily. So Porter was not the only one to have such thoughts.

Few of the computer trailblazers had time to break off from their work to make predictions. But some did, and were woefully inaccurate in forecasting how the world would take to their inventions. In the late 1940s a group of prominent British engineers was called in by the government to give their opinion on what would be the impact of these new-fangled machines that had just appeared. The experts scratched their heads a bit and concluded that computers would not lead to a great deal of change. There would be, they said, just enough work in Britain to require a total of five computers.

It was a similar story on the other side of the Atlantic. IBM, now the world's biggest computer firm with annual sales around the $17 billion mark, made less than twenty of one of its first models because it thought the world could not possibly cope with more. Some forecasters were a little more realistic. According to French computer engineer Jacques Maisonrouge, one group of American experts prophesied in 1955, when there were all of 250 computers in the US, that the country would contain 5000 of the machines in 1965.

Now let us forget the forecasts and look at what happened. By 1965 there were 20,000 computers in the US. The Americans had underestimated by a factor of four. These computers would have been what today we call mainframes or minicomputers, that is, reasonably big machines that take up the space of a wardrobe at least. Ten years later, in 1975, there were 80,000 such machines in the US, with roughly as many more in the rest of the world.

But the real growth in the popularity of the computer was still to come. It was fuelled by the advent of the microcomputer, that is, a computer which comprises a number of integrated circuits on one or more chips of silicon. From sales of virtually nothing in 1973, something like twenty million microcomputers were sold world wide in 1979. According to estimates in the computer industry, the yearly sales total will continue to increase until it reaches 100 million in 1983. The US alone, which will continue to be the biggest market for computers, could contain about 100 million of the 'micro' machines by 1990, that is, one for every two people.

When adding up computers, we must not forget the better established 'big' mini and mainframe machines which, of course, for many years constituted all the computers there were. In 1979 there were some 700,000 big machines in the world, the figure being compounded from several computer industry estimates and including all machines costing £15,000 or more. The number is very small compared with what could be called the 'new wave' microcomputers. But the mainframes and minicomputers are important not just because they represent the established order of computers and carry out a lot of crucial roles. They are constantly affected by microelectronics in that the technology that gave the world the microprocessor is making the big machines both smaller and cheaper too.

The mainframe was the first type of computer. All today's mainframes can trace their predecessors along a fairly direct family tree to the ENIAC. Mainframes are the biggest, most powerful types of computer. Science establishments and industry use them for the heavyweight jobs that a lesser machine would find impossible to tackle. Thus they can be found computing weather forecasts by taking into account huge volumes of climatic data, or doing other 'number crunching' jobs, such as solving complicated scientific equations. The power of mainframe computers is reflected in their price – which typically is anything between £50,000 and £300,000. There are instances, however, of mainframes costing several million pounds. One is the machine the US space agency NASA is building to work out highly complex design equations for aircraft.

If the mainframe is the heavyweight of the computer world, the minicomputer is the welterweight. These computers entered the scene in the early 1960s, about the time miniskirts and minicars were fashionable. Like these items, not only were the new computers smaller than their predecessors, there was something slightly risqué about them as well. They promised to knock some of the stuffiness out of computers and were living proof that people could own one without having a special room and a lorry-load of attachments to accommodate it. Minicomputers started the computer boom. They are generally about the size of an ordinary household refrigerator and cost somewhere between £15,000 and £30,000. They can be seen anywhere there is an office, a factory floor or a science laboratory for them to take up residence. In 1978 the world's minicomputer makers sold about 150,000 of their offspring, and by the mid 1980s this figure could rise to some 750,000. All told, there were about 500,000 minis in the world in 1979.

The final member of the computer family is the one that really made it into the limelight, the microcomputer. Microcomputers have carried on with the trend, started by the mini machines, of making computers acceptable – only at a greatly increased rate. The term microcomputer was coined, no doubt, because no one could think of describing machines smaller than a 'mini'. To be sure, microcomputers are normally pretty small, though usually not as tiny as a chip. They require attachments such as a keyboard, screen and printer and, with these, add up to approximately the size of a TV set. But in reality the boundary between mini and microcomputers is fast becoming blurred. Increasingly, the distinction between the two looks nonsensical. There are two reasons for this.

First, both kinds of equipment are being made out of lots of little chips linked together, with a mini machine only slightly more powerful than a micro. Second, the roles of the two types of machine are beginning to overlap. What a few years ago might have been a minicomputer tucked away in some office corner is today likely to be found sitting on an office manager's desk as a general office machine. A few buildings down the road, you might find almost exactly the same piece of equipment sitting on another desk and called a personal microcomputer.

A few years ago, people might have said that minis and micros will overlap but mainframes will always be a case apart. It could have been argued that mainframes would for years retain their big, unwieldy appearance because of the amount of power they contain. However, recent events upset even this point of view. In 1979 IBM announced what it called its 4300 series of mainframe computers. Amazingly, these machines, for all their great power, were not very big at all. In total a 4300 takes up about the space of a sideboard; but its central logic and memory unit is the size of a shoebox and contains just eighty chips. IBM used all the latest advances in microelectronic technology to cram a lot of power into a little space. And the result was a machine surprisingly cheap: it costs a mere £100,000.

There is yet more confusion. The latest microprocessors, just off designers' drawing boards, will rival even the power of the 4300 machine, so many in the computer industry suggest. Soon there will be no difference between the three types of machine.

Of course, the names do not really matter. Perhaps it would be better if they were phased out. What is important is that computers are becoming more and more unobtrusive. In more of the gadgets that will emerge from the engineering world over the next ten years, the computer components will be largely hidden beneath the familiar exterior of goods like cars, washing machines, toys, games and industrial machinery. In ten or twenty years people will barely mention computers, let alone write books about them. The computer, though its effect will be very much felt, will simply be invisible: perhaps, then, a better term for the emerging new breed of machine would be the 'vanishing' computer.

The computer is about to 'disappear', or more correctly become invisible, very quickly. After all, it has only been around as a workable device for a few decades. If the story of the computer were made into an Agatha Christie novel, the leading character would have vanished from view around the end of chapter one, leaving the rest of the characters to work out where on earth he had got to over the next two or three hundred pages.

Even though the history of computers is very short; to make sense of how they developed we need to go back a long time, to the time when people first felt the need to count.

A computer is an automatic, universal, calculating machine – it runs with the minimum of human intervention and it can be programmed to do many things – that makes logical deductions (decisions) on numbers, or simulations of events coded as numbers. Let us go back into history to discover when people first found numbers important.

To early mankind, counting was crucial. It meant our ancestors could quantify the important things in life. They could determine how many people are in the tribe, what is the distance between two villages, how many animals have just been killed in a day's hunting. People first learnt to count, we must assume, by drawing lines in the soil with a twig, or perhaps piling up pebbles on top of each other. Each line (or pebble) represented one of the objects that the person was counting. This system would soon become unbearably tiresome once big numbers were encountered. Imagine one of our ancestors trying to count a herd of say one thousand woolly mammoths by drawing lines in the ground – he would soon run out of space. So the next step was to find a convenient large number to which the quantity of objects being counted could be related. This large number is called a mathematical base. Numbers larger than the base are represented as so many multiples of it: numbers less are fractions of the base. The obvious choice was for a base of ten – the number of fingers on two hands. By counting up to his total of fingers and making a mark in the ground when he got to the final thumb to represent the number ten, our caveman ancestor would require only 100 lines to describe his herd of mammoths, a far more manageable proposition than 1000.

This kind of counting process, with the incorporation of 'sub-bases' of powers of 10 such as 100 and 1000 to help to describe very large numbers in our familiar 'column' arrangements, has given us our decimal system of counting, named after the Latin word *decimus* meaning tenth. The decimal system is so ingrained that it is difficult to think of any other method for counting in a general sense. But humans are creatures of habit; and we can look back only a few years to see how people in Britain coped very well with a base of twelve for the shillings and pence monetary system, that was eventually phased out to angry mutterings that 'we'll never get used to decimals'. Other common bases are 360 (the number of degrees in a circle or seconds in an hour) and 100 (now the base for Britain's coinage system and for most others elsewhere).

The fact that there are other bases besides ten makes us realise that if it were not for the anatomical fact that people have, and always have had, ten fingers we might have ended up with a different base for counting. Like a base of two for instance. To most people this seems absurd. We could get by with a base of two – after all we have two legs and two arms so logically it is no worse a system than the decimal format – but it would be very laborious. Every time a person got to the number two he would have to do the arithmetical equivalent of drawing a line in the ground, which is to add a new column to the row of figures representing what he is counting. As two is not a very big number, there would soon be a lot of rows unless the number being counted was pretty small too.

To see how you can count with the binary system, which is what the numerical system with a base of two is called, consider the number which in decimal we call 3. Represented as a series of marks in the soil, this would be III. In binary, the first two strokes add up to 2, which is the base, which we must find a way of denoting. We give this the figure 1 (there being one group of 'twos'), leaving the one mark left over to be represented by the figure 1, only in the column to the right of the first one, denoting this is for the units. Thus the number is 11; the decimal number 2 on the other hand is 10 because there are zero units. Other numbers larger than 3 are represented in a similar way, building up columns of figures. As in the decimal system we have columns for units, tens, hundreds, thousands and so on. The difference is that in the binary system the rows of figures correspond to 'units', 'twos', and then the higher powers of two, that is four, eight, sixteen and so on, whereas in decimal notation the rows represent the higher powers of ten. We can see pretty quickly that a hallmark of the binary system is that only two symbols, 0 and 1, are needed to describe any number; the upshot of this is that a number which in decimal is described in a few rows using any of ten different characters can be defined in binary with only two types of character but many more rows.

Every decimal number can be put into binary. Hence the binary equivalent of 0 is 0; the binary equivalent of 1 is 1. After this things start to get a little more difficult: 2 is equivalent to $(1 \times 2) + 0 = 10$; 3 is equivalent to $(1 \times 2) + (1 \times 1) = 11$ as we have already seen. Other numbers can be given their binary equivalents in a similar way:

Decimal	*Binary equivalent*
4	100
5	101
6	110
7	111
8	1000
9	1001
10	1010
11	1011
12	1100
13	1101
14	1110
15	1111
16	10000

Thus, five digits are required to define a number that in decimal only needs two. Try working out how many binary digits you need to describe the number one million. You would find that if people were to use binary arithmetic as the standard way of doing sums, we would need an awful lot of paper to have space to put all the columns.

This is not to say binary arithmetic is difficult. Just the opposite in fact. Thus addition and subtraction can be carried out in binary in a similar way as in decimal, only with just two digits to bother about and three rules to learn, that 0 + 0–0, 1 + 0–1, 1 + 1–10. By contrast, in the decimal system there are the ten digits 0 to 9 to remember, plus the results of fifty-five different additions from various combinations of the ‘family’ of decimal digits.

The binary system, then, has got some attractive features about it, even though humans do not find it such a good way of doing arithmetic as what is in many ways a more clumsy decimal system. For computers, however, binary is an almost perfect way of doing arithmetic. First, the fact that there are only two digits means they can be represented by two electrical states, the most convenient being when current is either on or off. This makes the binary system a good choice for any electrical set-up. Second, the simple rules governing binary addition and subtraction (and therefore multiplication and division as these are the same as either repeated addition or repeated subtraction) make the system useful for a fairly dumb instrument, such as an electrical computer that does not have the resources of the human brain to make deductions. Third, the fact that you need a lot of columns to do binary arithmetic does not worry a computer. In computer terms, a lot of columns simply translates into the need for a lot of electrical circuits which can be accommodated inside the machine very easily.

We are surging ahead rather and starting to touch on how a computer makes its decisions. This will be explained properly, later in this chapter – suffice to say for now that it finds the binary system of arithmetic very useful, which is why almost all modern computers use it.

Let us return to the caveman with his decimal system that was to set the tone for human counting for many centuries to come. The caveman looked around for something a little more sophisticated than using his fingers. The next development in counting after the ‘finger system’ was the abacus, truly the first computing machine. The abacus reached the peak of its popularity around 50 BC, when people all over the world used it; you can still find some men and women in places like Japan who would be lost without an abacus; and of course it is a standard piece of equipment in many nurseries the world over.

An abacus is a number of strings of beads on a wooden frame. A column of numbers is represented by the number of beads that are pushed to the bottoms of their respective strings. You can do simple arithmetic by changing the numbers of beads at the bottoms of the strings, with tens and hundreds being carried over to the next 'column' just as in addition and subtraction on paper. The more complex steps of multiplication and division are done with a lot more effort, by treating these processes as either repeated addition or repeated subtraction, as in the case of the finger system. However, the abacus was a step ahead of the fingers because with it people could deal with much larger numbers; and they had a system for 'storing' them (as a group of beads on a string) which they didn't have before, once they had run out of fingers that is.

It was not until the 17th century that the world found a way of simplifying the chore of doing multiplication and division. A Scotsman called John Napier invented the idea of logarithms, the name usually being shortened to logs. Napier discovered that associated with every number there was always another number called its logarithm. The value of this logarithm is the power to which another number, called the base, has to be raised to give the original number. We have come across the idea of a base once again and, no surprises, the base normally plumped for is our old, familiar one, of 10. You can see that working out logs for simple powers of ten is very easy. The logarithm of 10 is the power to which 10 has to be raised to give ten, which is 1. Similarly, the log of 100 is 2 and the log of 1000 is 3. All the numbers between these straightforward ones are more difficult to work out, but can be looked up in what are called log tables in which someone has gone to the bother of working them out for you. Numbers less than one have negative logs.

The real usefulness of logarithms, apart from introducing the idea that numbers could be put into the form of other numbers to make them easier to handle (this would be important once computers were introduced), was that you could use them to do multiplication and division. It could be said that the idea made these steps as easy as falling off a log.

It is a property of logs that if you add the logarithms of two numbers, the result is the logarithm of these numbers' product. Similarly, to divide two numbers you merely subtract their logarithms. The process made arithmetic an awful lot easier. It also gave birth to what became the engineers' and mathematicians' standard tool for three centuries until it was supplanted, in the early 1970s, by the hand calculator – the slide rule. A slide rule is a couple of pieces of wood like rulers linked together with numbers on them. (The explanation would not have been necessary a few years ago, but, due to the pocket calculator, slide rules have now taken on the status of historical objects and have largely disappeared.) By sliding up and down the bits of wood in relation to each other, you can effectively use logarithms rather more easily than by looking up the numbers in tables, and hence the usefulness of the instrument to anyone whose work involved calculations.

Now let us get to some definitions which are vital for understanding how a modern computer works. They concern the two computing machines (to use this term loosely) that we have just encountered, the abacus and the slide rule. Both are, fairly obviously, mechanical devices: there is nothing electronic about either of them so they are far removed from what we would think of as computers today. But in one respect the abacus is much nearer a computer than the slide rule.

The slide rule is termed an analogue machine. This is one which sets up a correspondence, or analogy, between what is to be measured or calculated and another quantity which allows the process to be carried out. In the slide rule, numbers are represented as logarithms which are themselves represented as lengths on a strip of wood. By making an analogy between a numerical quantity and a length, the slide rule helps a person carry out a calculation. This is in the same way that someone often makes an analogy in conversation to press a point home.

A good example of an analogue machine is the ordinary household clock. When you look at a clock-face, you are setting up a correspondence between the positions of the hands and the numbers which these positions represent. In this way, even if the numbers on the dial are missing (as they often are) people know that if both hands are upright, it is 12 o'clock.

Why is a slide rule fundamentally different from a modern computer? The answer comes by again considering the abacus. Unlike a person with a slide rule, someone using an abacus is concerned with numbers in a purely abstract sense. He does not relate them to anything else in order to make it easier to do the calculation. To be sure, the person needs the beads to be able to count, just as a caveman watching mammoths needs his fingers or marks on the ground. But the marks could just as easily be bars of chocolate, or cigarette ends, or rusty nails, or anything else for that matter, and the person could still do his calculation just as easily. The abacus is concerned with counting. The slide rule is concerned with measuring. Because of its dependence on counting numbers or digits, the abacus is called a digital machine.

For our purposes, all modern computers can be considered as digital machines which, furthermore, use the binary system of arithmetic. There were indeed some early analogue computers, most of them mechanical systems (although electronic analogue machines were known). They again did their calculations by representing numbers as measurements. Such machines incorporated elaborate systems of wheels and axles. These components turned, or otherwise moved position, to represent certain mathematical processes, the results following by someone measuring the distance a pointer has moved or a wheel has revolved.

These mechanical machines were extremely ponderous, however. They were largely swept away in the electronic revolution that came in the 1940s.

To go back to the digital machines, those that followed the abacus worked in decimal fashion, not binary. It took a long time for people to realise that what was the best system for humans was not necessarily the best for machines. The calculating machines of the 17th and 18th centuries were mechanical contraptions that essentially used wheels and axles to accomplish what had been done on the abacus. Depending on how these components intermeshed with each other, stops were knocked over which signified a new number. By altering the system in some way and setting different numbers into the machine through putting the stops in different positions, you could add or subtract, and, by repeating these processes many times (this was fairly quick on account of the mechanics of the system), multiply or divide. Look at how your gas meter counts the number of gas units your home consumes if you want to learn how such a counting mechanism works.

There was still one big problem with these digital counting machines. They required constant attention. If you wanted to change the equipment to do a new calculation, or to act on a different set of numbers, you had to keep fiddling about with it, writing down the results of the last set of calculations while doing so. Was there such a thing as an automatic calculating machine?

There was. An eccentric professor of mathematics at Cambridge University called Charles Babbage designed one in 1812 which he called the difference engine. A few years later, in 1833, he went one better. He designed a calculating machine that was both automatic, that is, did not need attention, and universal, that is, could be put to a variety of uses – which is pretty well our definition for a computer and which sets it apart from a calculating machine. It is for this reason that Charles Babbage is called the father of the computer.

Unfortunately, neither the difference engine nor the analytical engine of 1833 were ever built. Babbage was far ahead of his time, so far that few people could follow what he was on about. Most crucially for him, the craftsmen of the day could understand hardly anything of his drawings. Even if they could have understood them, it is doubtful whether the engineering skills of the day were high enough to produce Babbage's strange pieces of equipment. Even worse than this was that Babbage kept changing his mind about what he thought he wanted: when he died, an unhappy man, the local engineers must have been quite relieved to have the old fellow off their backs.

Babbage's day was to come, however, for he gave the ground rules for computer design which are still basically in force today. His analytical engine was a logical development of the geared systems of the earlier calculators, only it contained a mechanism for linking the results of one arithmetic step to the next operation. Also there was a way of controlling how the machine functioned which could be altered fairly easily – this gave the world the idea that people could control a system with an easily changed software program, or set of instructions.

The professor's machine consisted of:
a) the arithmetic unit which carried out the basic calculations,
b) the store (now known as the memory) which holds numbers before and after they are acted upon in the arithmetic unit,
c) the control unit which ensures the arithmetic unit and the store interrelate in the way the user wants,
d) the input device by which numbers are supplied to the machine and
e) the output device by which the operator receives results.

These five parts are the basic bits that are in most computers today, from mainframes to microcomputers.

In Babbage's design the memory and arithmetic unit were wheels and gears of the kind seen in previous calculators. Transfer of numbers from one unit to another was by a system of rods and other kinds of mechanical link. Output and input devices were the series of stops and so on already described. What was radical about Babbage's proposal was the method of controlling how the machine operates. This was to be done with a series of punched cards, which another inventor, Joseph Jacquard, had already devised for controlling a loom in weaving. Sets of cards with holes punched in them are made to run over rods and wheels which move in a prescribed pattern according to where the holes are. You can see the same kind of approach in old-fashioned fairground steam organs where the forcing of steam through the organ pipes is controlled by a sequence of punched cards. And in many of the electronic computers of the 1960s and the early 1970s, punched cards still did the same kind of job as in the analytical engine, only working with an electronic system instead of a mechanical one.

Babbage's punched cards were the first computer programs. By changing the cards, or inserting a new set of holes in the machine, the computer could be made to operate differently. Something similar happens in the case of a fairground organ or a modern gramophone system where changing the 'software' (another set of punched cards or a new record) makes the machine play a new tune.

Various people made tabulators and other equipment that owed something to Babbage's ideas. But no one went much further than him until an American engineer called Vannevar Bush arrived on the scene in the mid 1920s. He built what was called a differential analyser. This was an analogue computing machine because it represented mathematical processes by a series of gear and wheel movements, the machine's theory having been worked out by the British Lord Kelvin in 1876. British engineers heard of the device, went to visit Bush in America and built a replica of the machine just before the Second World War at the University of Manchester. Among the engineers was a young undergraduate named Arthur Porter, who we met briefly at the beginning of this chapter. He was assisting one of the most eminent men of early computing, Manchester professor Douglas Hartree, who used Meccano parts to build a model of the machine before going ahead with the full version. This episode was recorded in a pleasant note in the July 1934 issue of *Meccano Magazine*: 'Meccano had provided him [Hartree] with many happy hours when he was a boy and had been of great assistance to him on many occasions. Naturally, therefore, he used Meccano parts for the building of his machine.'

This warm message probably helped sell a few more Meccano kits to aspiring youngsters over the following few months. Today it gives a fair indication of the amateur enthusiasts' spirit in which the early computers were built. The first computer builders could easily have come out of the pages of *Boy's Own*. The coming of war changed all this. War concentrates people's minds wonderfully. It certainly did something for speeding up the development of computers. Hostile governments suddenly had a reason for making machines that would work out complex calculations at high speeds – problems such as the trajectories of missiles, how to decode enemy messages, what to do, even, with certain radioactive and highly unstable materials if you wanted to make a nuclear bomb.

The war changed the course of computer development, and made engineers alter their ways drastically. Meccano parts went out of the window for good. With them went the idea of mechanical computers, whether analogue or digital. Engineers turned instead to simulating arithmetic steps with electrons flashing along wires or across evacuated tubes called valves. In a mechanical system, a three-digit number is represented by three gear wheels. In an electronic system, the same number can be defined by a group of ten valves, each of which characterises one of the digits 0 to 9. Electric pulses, passed through the different valves in turn, give the electronic representation of a multi-digit number. The machine can perform addition and subtraction by electrical interaction of the different pulse trains in a way analagous to the linking of wheels and axles in a mechanical machine.

The first tangible result of this line of work was the ENIAC (it stands for electronic numerical integrator and calculator) which engineers at the University of Pennsylvania built in 1945. The ENIAC is normally thought of as the world's first electronic computer, although British machines called the Colossus series, built like ENIAC as part of the war effort under utmost secrecy, were completed around the same time.

Notice that these machines, though electronic and digital, still worked by the decimal system. They needed ten rows of tubes to define every number. The next, decisive step in the history of computers came just after the war finished. An engineer at the University of Pennsylvania, John von Neumann, gave a series of lectures on the binary system and its potential for computers. He explained that not only was a two-digit system admirably well suited to an electronic device that could recognise current 'on' or current 'off' as equivalent to either of the two digits, but that the binary system could give computers a way of working out *ideas* not just numbers. Von Neumann used a logic code worked out a century before by the English mathematician George Boole. He said that the binary system could act, not only as a code for the decimal numbers familiar to everyone, but for thoughts and actions which were part of everyday life. This was a bombshell. Until then mathematicians had looked upon computers as increasingly more advanced machines for performing numerical calculations. Now someone was proposing a far wider use which had virtually no limits. Von Neumann's ideas, and those of others like him, took a while to work through. But they set the scene for the huge range of applications possible for today's microprocessors, and the idea of the all-encompassing 'decision-making machine'.

Electronically, a binary digit is represented as either current 'on' or current 'off' – a similar system could be low voltage or high voltage. There are only two binary digits to worry about, 0 and 1, so 0 is made to correspond to current 'off' and 1 to current 'on' (or 0 to low voltage, 1 to high voltage depending on what is most suitable). Such an arrangement is far simpler than in the ENIAC for instance, where ten different electrical states are needed to represent numbers. To define a particular number comprising several binary digits, a sequence of 'ons' and 'offs' would be needed. Thus the binary number 5 (which is written 101) is represented by a 'current on, current off, current on' sequence in three separate circuits. Each of these circuits acts like switches, in principle similar to the switches that turn light bulbs on and off in rooms. A set of n switches or circuits gives the possibility of 2^n – different patterns of binary digits (or 'bits' as they are normally referred to).

A set of switches or circuits can represent words as well as numbers. Each letter of the alphabet is represented, under an international code, by a pattern of eight bits. So eight circuits, multiplied by the numbers of letters in the word, is enough to define any word. Under this system the word 'door' can be defined by thirty-two circuits but 'table' would need forty. In this way we can see how a computer can store electronically the contents of books, magazine articles and so on. All the words that these publications contain are broken down into their bit codes, then stored as electrical signals of current 'on' or 'off'. A medium-sized novel of 100,000 words, each containing, say, five letters, could be stored in about four million circuits.

This, then, shows how a computer can store numbers, or anything coded in numerical form such as words or letters. All you need is enough wires or some other medium for electricity that can carry current or bursts of electrons. Very crudely, this is how the stores, or memories, of both the first old-fashioned computers and today's microcomputers work. But how does the computer make the numbers interact to do its calculations or 'make decisions'? This is where we start to consider what goes on in a computer's central processor, which is more or less equivalent to the arithmetic unit in Babbage's machine.

The central processor does the electronic equivalent of making deductions. Under the guidelines laid down by von Neumann, who himself borrowed much of the mathematical theory from the Englishman Boole, computer designers use logic patterns borrowed from ordinary life to make their machines work.

Logic is a way of reasoning that uses the two statements 'true' and 'false' to work out the meaning of particular sets of conditions. Anyone who has ever played Twenty Questions has an inkling of how the system works. In this game certain statements are labelled either true or false, and with these labels it is usually possible to discover the identity of a 'mystery' object. Logic can be applied not only to identify things, but to decide on a course of action. Consider the following three statements about, for the sake of an example, a cat:

1) the cat is behaving itself
2) the cat is ready for its dinner
3) the cat is dead

If the first two statements are true, and the last one false, a considerate cat owner would set about preparing the animal's dinner. If the first and last statements were false, the person might do nothing, just to teach the cat that it would only get its food when it stopped being naughty. On the other hand, if the last statement were true, the owner would realise he had better start digging a hole in his garden.

In a computer, an electronic statement corresponding to the binary digit 0 (the statement is usually 'no current') is equivalent to the statement 'false'. The binary digit 1 corresponds to 'true'. Thus deductions like the ones I have just made in the case of the cat could equally well follow by flashing 'current on' or 'current off' signals through a set of three circuits each referring to one of the conditions (1), (2), or (3).

A computer can cope with whole series of logic conditions, each one of which is either true or false, finally arriving at an answer or a 'decision', with a series of logic circuits known as gates. There are three basic types of gate called the AND, OR and NOT gates. (Don't worry about the names – just remember they are very logical.) The AND gate produces a flow of electricity (corresponding to the digit 1 or the statement 'true') only if the input circuits to it all carry similar pulses. Thus it will provide an 'answer' of 'true' only if all the inputs have the tag 'true' associated with them as well. On the other hand, the OR gate produces an answer true if at least one of the inputs are true. The final gate, the NOT, is an inverter that changes true into false and vice versa.

We can go back to the example of the cat to show how these gates work. The input circuits to our gate could represent the conditions 'the cat is behaving itself' and 'the cat is ready for its dinner'. The output circuit would represent 'the cat must be fed'. If the gate was an AND gate, the output would show 'true' only if *both* input conditions were true; if it was an OR gate the decision would be to feed the cat if it was either ready for its dinner *or* behaving properly.

Examples about cats are not perhaps all that realistic because there are not many computers programmed to tell pet-owners when to feed their animals. (There is, however, no earthly reason why this should stop anyone who thinks it is a good idea). But the examples show how chains of circuits transmitting binary digits as either a pulse of electricity or a gap of 'no pulse' can make deductions. In a chemical process in a factory, for instance, circuits could be positioned that transmit 1s or 0s depending on whether certain things happen, such as if a temperature reaches a certain level or if the surface of a liquid climbs to a particular height. Depending on how these 0s and 1s are processed by AND, OR and NOT gates, the computer could decide certain things, for example, whether to turn off a heater current or to stop the flow of a reagent into the process. The complicated decisions that powerful computers are capable of producing require thousands of gates. Even a relatively simple circuit that will add binary numbers requires nineteen gates, eleven of them AND, five ORs and three NOTs. In practice, it would obviously be ludicrous to have to change the gate structure of a computer every time the user wants to do a different job with it. This is where programming becomes important. Putting a new program in a computer effectively changes the order in which its gates do their job and is the key to a computer's flexibility. Programming, or devising new software for computers, is far from simple – and can be very expensive. We'll discuss software problems in Chapter 4.

It is all very well talking about what these gates do but, so far, we have not explained what they are. In the early days of computers they were made up of valves (also called vacuum tubes) joined by wire to resistors. A valve is an evacuated glass tube through which a current can pass from a heated negative electrode (called a cathode) to a positive electrode (an anode) at the other end. In the middle, between these two electrodes, is another called the grid which can be maintained at different voltage states to influence whether electrons pass through it. This happens because the grid has holes in it like a sieve. Small changes in the voltage applied at the grid influence whether electrons flow through the valve between cathode and anode, so constituting current. In this way the valve acts as a switch useful for putting in an AND, OR, or NOT gate. Simplifying somewhat and representing the valve as a straightforward on/off switch, AND, OR and NOT gates appear as in figure 1.

As this is not a physics text book, I do not propose to enter any further into explaining how the first computer designers made their central processors. It is enough to say they consisted of hundreds, if not thousands, of valves joined up to other standard electronic components such as resistors and capacitors. Electrons flow through the various gates made up by these components, being switched first in one direction then another. The electrons act rather like a motorist who is constantly being shunted from one forked road to another by a series of extremely complicated diversions. If you think of electronic gates acting similar to gates in real life, either swinging open or remaining shut depending on the behaviour of the 'motorist' electron (for example according to whether or not he pays the

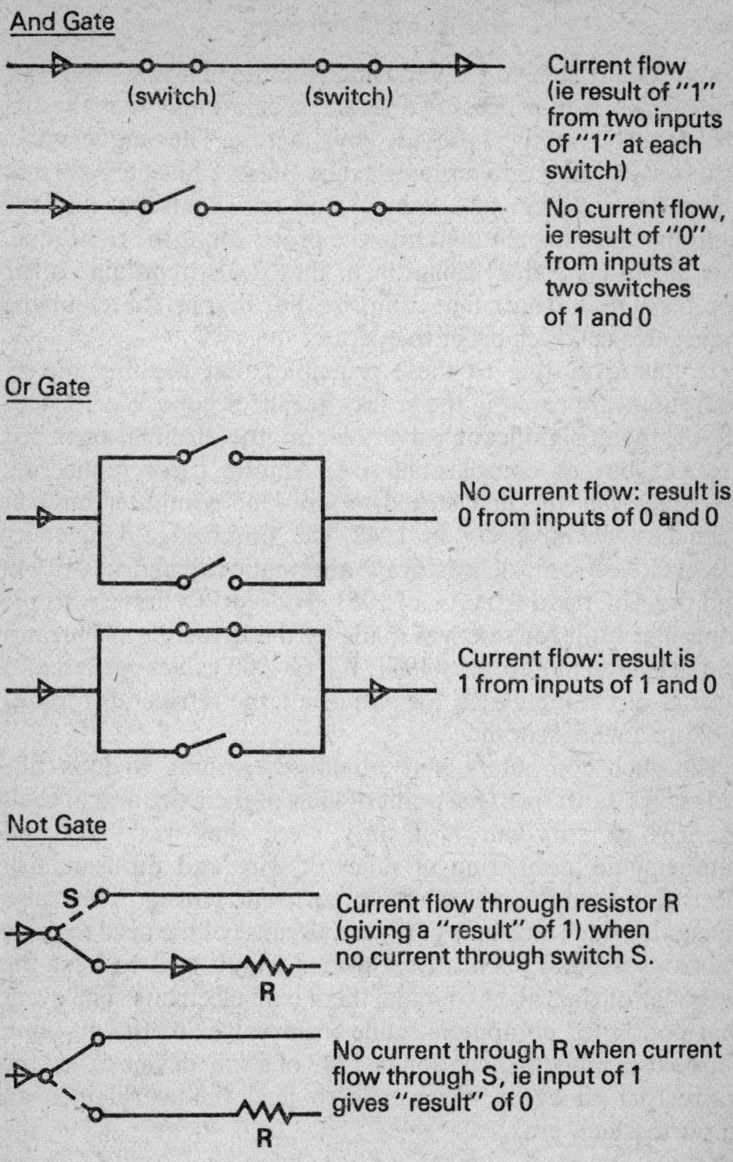

Fig. 1 And/Or/ And Not Gates in Computer Circuits

required toll) then you have a rough picture of how a processor unit works. The processor does its calculations (or makes its decisions) on numbers provided by the store. The numbers flow into it as binary code represented by pulses of electric current under the direction of a control unit which ensures that the right numbers are shunted into the processor at the right time. The store has a dual function in that it also contains other chunks of digital data that comprise the program for telling the logic gates in which order to operate.

It was according to these principles that the first binary computers appeared in the years after the Second World War, all of them significant advances on the decimal-operated ENIAC, but of comparable size. Among these pioneering machines was the first stored-programme computer built at Manchester University in 1948, the Cambridge University EDSAC (electronic delay storage automatic calculator) of 1949 and the American EDVAC of 1951. The world's first electronic computer built for sale was made by the British firm Ferranti and delivered in February 1951. It had 4000 valves packed into a series of cabinets each the size of a large refrigerator and it took up a whole room.

But such computers were already beginning to look like dinosaurs, both from the point of view of their size and of their chances of survival. Not only were they very big and cumbersome, consisting of miles of wire and thousands of electronic components which often went wrong. They also consumed huge amounts of power, because of the need to drive electrons around such an enormous circuit and to heat the cathodes of the valves to make them emit electrons. The event that consigned computers made from valves to the museum cupboards was the invention in 1947 of a tiny device called the transistor, an event which was to lead the world into an important new era.

Chapter 3: The 'brainbox' shrinks – the integrated circuit

The invention of the transistor set the world off towards a future of cheap, tiny electronic components. The world was never the same again. The manufacture of the first transistor was one of those important events that rank with the times when people first learnt to make axes out of stone, or drinking vessels out of bronze, or, more up to date, when they found how to harness steam power effectively and sparked off the industrial revolution in England some 200 years ago.

The transistor is a piece of material known as a semiconductor which, with its packaging, measured two or three cubic centimetres when it first appeared from production lines in the early 1950s. It did practically the same job as a vacuum tube taking up the space of some thirty cubic centimetres, a tenfold saving in space. Further, it was much cheaper to make and used a lot less power. How a transistor works will be explained later. But the important thing to grasp is that virtually overnight it shifted people's thinking about electricity – or how to harness the power of electrons – into a new dimension, the atomic dimension. Instead of constructing electronic devices out of tubes not that far removed from light bulbs, where the electrons travel several centimetres to do their job, engineers could make equivalent components from little globules of semiconductor material, in which tiny displacements of electrons are all that are needed to make the device work.

What brought the term 'microelectronics' into being was the fact that this 'shrinking' did not stop with the transistor. The transistor introduced the idea that semiconductor physics could be useful to electronic engineers who want to make small components – it was the breakthrough, if you like. But afterwards, the process of making transistors and joining them to other components was constantly refined so that they could fit into a smaller and smaller space. The result was that by 1958 or so a small piece of semiconductor about 5 mm square and 1 mm deep (called a chip) was all that was needed to contain one transistor. This represented a hundredfold reduction in volume on the first transistors. The chip did the job of a valve one thousand times bigger. After this the refinements in technology came thick and fast. In 1980, you could buy prototypes of devices that have 250,000 transistor-equivalents on a chip-sized piece of semiconductor. If you were able to separate just one transistor on this chip from all its fellows (which in fact you can't) you would find it as useful as a glass tube an astonishing 250 million times bigger. This, then, put the magnitude of the size reduction that has come about in microelectronics in its full, amazing, perspective.

Now some words of advice. The fact that this 'shrinking' has happened and that lots of tiny transistors, with a few other electronic odds-and-ends thrown in for good measure, add up to one pretty small computer is all that you basically need to know about microelectronics. Armed with this knowledge, and the awareness that making a small computer may be easy but making it work can be a lot harder, you can face the world confident that you are not entirely ignorant of the new forces that are making it tick. If you are happy enough with this state of affairs, miss out the rest of this chapter and proceed to the next one. But if, on the other hand, you are curious to know just how so many devices can be crammed into such a tiny space, and what makes them work, then the rest of this chapter should enlighten you.

A group of engineers at Bell Laboratories in the US invented the transistor when they were experimenting with the class of materials called 'semiconductors'. The most common semiconductors are germanium and silicon. Germanium was the most important in early transistors although silicon, a very abundant element which is found in sand, is used in nearly all of today's microelectronic devices. Semiconductors are strange materials compared with most of the others on Earth. They are peculiar because of their electrical properties. All materials contain atoms which have a certain number of electrons clustered round them. We already know that the movement of electrons constitutes electricity. Now most materials are either very good at allowing their electrons to move through them to a source of positive voltage (electrons move towards a positive potential and away from a negative one because they are negatively charged and follow the old rule that 'like charges repel, unlike ones attract'), or else they are very bad at letting their electrons move. Materials in these categories are known respectively as conductors or insulators. Examples of conductors are most metals – this is why metals such as aluminium or copper are made into electrical wires – while common insulators include plastics, which are often used to prevent electricity reaching places where it is not wanted.

Semiconductors, however, are midway between these two types of material in their electrical behaviour. While metals conduct electricity by a great splurge of electrons flowing from one point to another, and insulators chastely refuse to allow anything of the sort, semiconductors conduct or don't conduct according to the circumstances. You can control the movement of electrons through semiconductors with precision by making slight adjustments to the electrical forces that act on them. And this is of great importance in making the materials into useful devices.

We already know from the last chapter that valves, evacuated glass tubes about 15 cm long containing three electrodes, can act as switches. In this they are useful in the logic gates and the memories of computers. How can pieces of semiconductor do the same?

First, we must understand that most semiconductors need their ability to conduct electrons beefed up just a little before they can be useful. This is done by adding impurity atoms to the semiconductor crystal. The impurities either have a few 'spare' electrons of their own, which they donate to the electron pool in the crystal and so make it more easy for a current to flow. Or, alternatively, they have a deficiency in electrons which, though it might seem hard to appreciate, is just as useful.

When impurity atoms with extra electrons – such atoms include phosphorus and arsenic – are added to the crystal, conduction takes place by the movement of negative charges or electrons. Hence such 'doped' semiconductors (as impure semiconductors are called) are known as n-type materials. There are even fewer electrons to go round when atoms such as boron, which already have a lack of electrons for their own purposes, are added to the semiconductor crystal. But conduction can still take place, not by electron flow but by movement of electron vacancies, or simply 'holes'. Holes don't actually exist as physical entities but it is necessary to invent them to explain semiconductors properly. Think of holes in the same light as the empty spaces on a draught board with all the other spaces taken up by draughts, which, under this analogy, become the equivalent of electrons. As a draught advances one square to take over an empty space, leaving a vacancy behind, a 'draught space' or hole may be thought of as moving the other way. Thus it is that, in semiconductors, holes move in the opposite direction to electrons, towards a source of negative potential as though they were carrying a positive charge. For this reason, semiconductors doped with hole-providing atoms are known as p-materials.

The difference between n- and p-materials is important in microelectronics. If you were to look very closely at today's microelectronic chips, you would see thousands upon thousands of 'islands' of n- and p-materials set into the 'sea' of the silicon substrate. That is anticipating the future, however, because we are still at the stage of considering the first, simple transistor – which in fact consisted of just three 'islands' of n- or p-material joined together.

William Shockley, Walter Brattain and John Bardeen, the three Bell Laboratories' employees normally credited with the transistor's invention, were connecting up n- and p-materials when they made their discovery. Let's look at what happens when you connect an n- and a p- region 'back to back' in an electrical circuit, remembering that an n-material conducts electrons towards a positive charge, and a p-region permits the passage of holes towards a negative charge.

In a circuit comprising a battery and a pn junction, as this arrangement is called, there are two possibilities – either the positive terminal of the battery is connected to the p-region and the negative terminal to the n-region, or the other way round. In the first instance, holes in the p-region are attracted across the pn barrier to the source of negative potential. At the same time, electrons from the n-region travel the other way to the positive charge. Hence there is a flow of current.

If you connect up the battery the other way round, however, electrons from the n-region and holes from the p-domain are attracted away from the barrier and no current flows around the circuit as a whole. The pn junction in the circuit is therefore fussy over whether it lets current flow. This happens only when the battery is positive with respect to the p-region and negative with respect to the n-region.

Figures 2a and 2b summarise how a pn junction works:

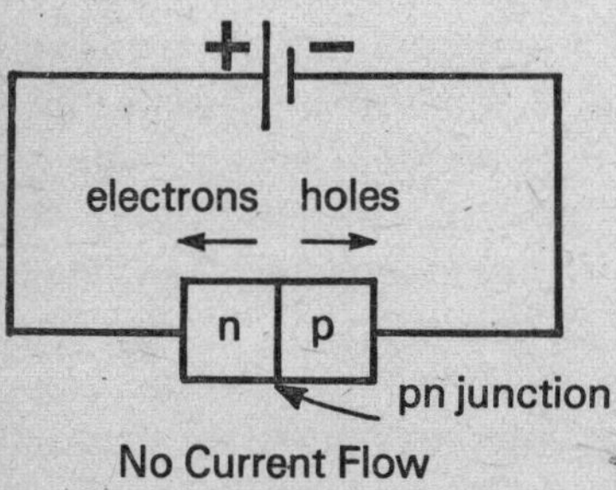

Fig. 2A

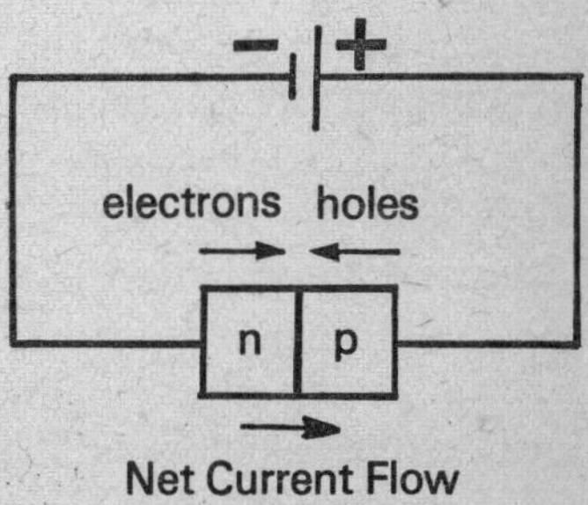

Fig. 2B

A pn junction can act as a kind of switch because current flows through the device only if the polarity of the battery is right.

But a much more useful device follows if you add a further 'island' of material, making a sandwich structure that is either npn or pnp in formation. This is what is known as the junction transistor.

We'll consider here what happens in an npn configuration (the pnp format works similarly). The 'n' parts of the transistor are known as the collector and the emitter, and the p 'filling' in between them is called the base. In a working transistor, the base is kept at a slightly higher potential than the emitter (that is, it has a slightly more positive voltage) and the collector in turn has a positive potential relative to the base. The circuit appears as in figure 3.

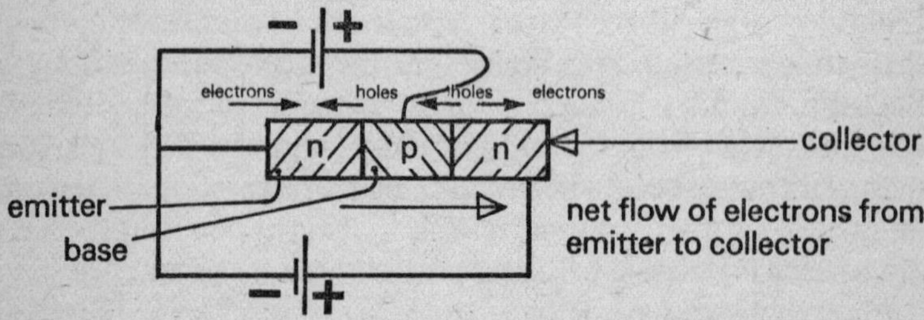

Fig. 3

At the junction between the base and the emitter, electrons and holes cross from one region to another in the same way as they do in fig. 2b. This is because the n-region is at a negative potential, the p-region at a positive potential. At the base/collector junction, however, things are different. There the configuration is as in fig. 2a because the polarities are 'wrong' for conduction and no current flows. Nonetheless, because the collector is at a higher voltage than the emitter, and providing that the base region is very thin, most of the electrons that flow into the base are pulled into the collector region. Thus there is a net flow of current from emitter to collector.

The transistor's role as a switch comes about because unless the base voltage is positive with respect to the emitter, there will be no flow of electrons from the emitter to the base and hence no current through the device. A high or low voltage at the base thus influences whether the emitter/collector current is on or off. We can now see how a signal that represents the binary digit 1 (a high voltage) at the base produces another signal corresponding to 1 (a pulse of current) which flows from the collector. In this way circuits with transistors in them can act as switches that are on or off in the same way as valves, and can fit into the same kind of AND, OR and NOT gates that we met in Chapter 2. Transistors can also be put into circuits in computer memories: by controlling the base voltage, digits 0 or 1 can be shunted in and out to represent ordinary decimal numbers or groups of words. Like valves, transistors have three points on them that are electrically active, the collector corresponding to the valve's anode, the emitter to the cathode and the base to the grid. In fact, if you don't bother yourself with the details of how they work, you can consider a transistor as doing the same job as the valve with the electrodes in both performing almost exactly equivalent functions.

The last few pages contain the deepest excursions in the book into electronics theory – but rest assured there will be very little more of this theory. However, the explanation of how a transistor works is vital because transistors make microelectronics possible. A transistor consists of just three islands of material 'doped' in a certain way with impurities. It does not really matter how small these islands are – they will still work. It follows, therefore, that if engineers can make them very small, linking each set of islands to an adjoining set through the semiconductor substrate, and devising a way of putting the semiconductor equivalents of other components such as resistors and capacitors on to the substrate as well, then they can build up whole circuits called integrated circuits on to a tiny piece of silicon. And this is exactly what electronic engineers, clever people as they are, have done.

When you are thinking about this, try not to get carried

away. Put yourself in the position of the engineers of the late 1940s and early 1950s when they read of the invention of these funny little devices called transistors. They didn't exactly rush out and start building computers with them. For a start transistors were not very reliable and it was some years before factories could make them successfully in large quantities off production lines. Secondly, computers were fairly new entities, and the world was hardly begging engineers to make them with these small and cheap devices. The world had yet to find uses for computers – even though, as we can say with hindsight now, the uses were there staring people in the face. Computers had an uncertain welcome because they were strangers. In the words of Jack Howlett, a pioneering British computer engineer: 'Even the early machines were a thousand times faster than anything that had gone before. They were incredible achievements and you can't blame people for not really knowing what to do with them.'

The first markets for transistors were in replacing valve amplifiers in radios and hearing aids. These were devices that had been around a bit and which people had no hesitation in saying would be more useful if made smaller and cheaper. (Transistors replicate valves in another way in that both are good at boosting electrical signals in amplifier circuits.)

Gradually computer engineers realised that transistors could be useful to them. The first computer made with transistors was a Ferranti machine that saw the light of day in 1953. However, transistors in computers didn't really take off until the 1960s. They had to wait for production methods to improve. The first transistors were made with an alloy method in which the tiny bits of n- and p-material making up the sandwich arrangements are fused together. The area of silicon involved was about 4 or 5 sq mm. But this is a fairly clumsy method and so the filling of the base region could not be less than about 25 micrometres wide. (One micrometre, written μm, is a millionth of metre – by contrast the diameter of a hair is 20 to 100 μm.) Now the width of the base is crucial: it has to be thin enough for electrons to be drawn across it to permit emitter–collector flow. The quicker electrons can travel across this region, the faster the circuit reacts to outside events and the more useful it becomes. Transistor designers therefore wanted to make base distances as small as possible.

The technique that gave them the freedom to do this was devised during the 1950s. Instead of making each transistor separately, electronic engineers made a large number in one process that involved very small distances. The engineer starts with a large piece of silicon of about 10 cm diameter. He diffuses into it controlled quantities of the appropriate atoms to build up the different layers of n- and p-materials. The diffusion happens by heating the silicon to about 1000°C and allowing the dopant material, which at this temperature is gaseous, to infiltrate the silicon's lattice structure at the correct points. The process can be thought of as similar to a bit of blotting paper soaking up ink. By shielding those areas of the silicon he does not want doped with layers of impervious silicon dioxide, the engineer can make several hundred transistors in one operation. The transistors are then separated by dissection, packaged and shipped off to their happy owners.

It was not until 1959 that the integrated circuit was born. It seems now an obvious idea. Instead of cutting up the silicon wafer after the diffusion process is finished, why not link the individual transistors with ultra-thin metallic contacts to build up a whole set of components on one piece of silicon? This is exactly what the engineers did. People normally give the credit for the invention of the technique, called the planar process, to workers at one of the big US electronics companies, Fairchild. From this point, there was no turning back. Engineers found they could simulate the properties of other electronic components besides transistors on the silicon substrate or 'chip'. Thus the resistor (which is sometimes needed in building computers' logic gates) is a thin piece of n-doped silicon surrounded by a piece of p-material, or vice-versa. And a capacitor (needed for storing charge in semiconductor memories) is a piece of metal separated by an insulator, normally silicon dioxide, from another layer of silicon. By a gradual process of controlling more accurately the quantities of dopant diffused into the material, and defining more closely exactly where they went, electronic engineers found it wasn't too difficult to build up more and more transistors, resistors and capacitors on to a single chip of silicon some 5 mm square.

In 1964, the world digested what turned out to be a shrewd piece of forecasting now known as Moore's Law. The law has nothing to do with *Old Moore's Almanac* and turned out a lot more accurate than most of the statements that august publication contains. Gordon Moore, who was then director of research at Fairchild, suggested that the number of components that could be fitted on to a single chip would double every year. From 1959, the year of one component on a chip, to 1979, when it was possible to make one containing 250,000 components, Moore's prophesy turned out to be almost precisely in step with what happened.

Logic circuits made from junction transistors and resistors were not the only circuits to benefit from the advent of component integration. Gradually the field-effect transistor, or the n-MOS transistor, became more important as the methods of packing increased numbers of components on to a chip became more sophisticated. The n-MOS transistor is so named because it works by the movement of negative charge carriers (electrons to you or me) and is made up of three main components, a layer of metal, followed by one of oxide (silicon dioxide to be precise) and one of semiconductor (normally silicon). If you ever wondered how Britain's state-backed electronics company, Inmos, got its unusual name, now you know: it is adapted from the semiconductor process by which the company's founders hope to make their fortunes.

We will not bother going into the working of the n-MOS transistor except to say that, like the junction transistor, it contains three 'ports' or inlets, called the source, the drain and the gate. More important is that its method of construction requires fewer blocks or n- and p- 'islands' than for the junction transistor, so it takes up about a quarter of the space of the latter's chip equivalent. This is the reason that most high-density chips (for example memories and microprocessors) are not made of junction transistors integrated on to the chip, but of their n-MOS counterparts. The junction transistor, however, is faster than an n-MOS device because the base width, which influences speed, can be made very thin indeed, a fraction of a micrometre. This is why chips made from junction transistor technology (often called 'bipolar' chips) are seen in most big computers' central processing units, where speed, not the ability to pack components very tightly, is the governing factor.

Besides the chips made almost totally from transistors (of either of the two kinds) which offer logic functions, there is another very important class of integrated circuit. This is the memory chip. It carries out the same function as the store in Babbage's original design for his analytical engine. A memory chip is less complex than a logic chip: it stores packets of charge in lots of little capacitors. The packets represent letters, numbers of statements, coded as either 1s or 0s, or, in electrical terminology, charge or no charge. A capacitor in ordinary non-miniaturised electronics is a device that stores a quantity of charge, that is, a number of electrons, on a couple of conductor plates separated by a layer of insulating material. Capacitance is 'charge hoarding': it can be contrasted to 'charge leaking' which is when a current flows. An example of charge hoarding is the static electricity you feel if you touch a metal rail in a dry, centrally-heated atmosphere such as in large shops: the charge has built up on the metal and discharges when it finds a suitable outlet, for instance, your body.

In a memory chip, each memory cell constitutes one microelectronic version of a capacitor, plus up to four or so transistors which relay the charge to it. (The memory cells that take up the smallest space use only one transistor.) The main processor of the computer 'reads' data from the memory by arranging for the separate charges in each cell to leak out and travel through the computer's logic units. The processor 'writes' data into the memory by making the process happen in reverse. Similarly, the computer operator can inject fresh data into the memory or unload unwanted data either by injecting more packets of charge or taking old, unwanted ones out.

Nowadays it is just about possible to put some 250,000 memory cells (each containing one transistor and one capacitor) on to a chip half a centimetre square. In computer jargon, this would be called a 256K RAM, that is a random-access memory (a memory where it is possible to get almost instant access to any bit of data that it contains) with 256,000 storage spaces. Other devices in the RAM family include the 256K RAM's smaller brother (in memory capacity) the 64K RAM, and the really small fry, the 16K and the 4K RAMs. The bigger memories have successively become the standard in the computer industry as, year by year, the number of components that it is possible to put on a chip increases.

Chip memories are very important in the world of computers. Almost by definition, a computer is only as good as the amount of information that can be fed into it. With the latest techniques of large-scale integration, as the processes for putting lots of components on to a chip are called, a computer designer can store in a tiny space vast quantities of data as 'food' for his computer, and this greatly enhances its powers. A big computer might contain several thousand memory chips which between them could store perhaps fifty million bits of data. These feed into perhaps 100,000 logic gates in the central processor.

The random-access memory and other kinds of memory such as the read-only memory (where data can only be read not written), constitute the bread-and-butter of the semiconductor industry. About 300 million of these memory devices, the cheapest ones costing just a few dollars each, were sold in 1979. This is more than any other kind of densely-packed integrated circuit. (Standard logic circuits for big computers, which are not all that densely-packed, sold at the rate of about 3 billion yearly in the computer industry of the late 1970s.)

The microprocessor, the chip you tend to hear most about, will probably start outselling the memory chips in the mid-1980s. In 1979, about 30 million microprocessor chips were sold. A 'micro' contains the basic guts of a computer inscribed as a series of n- and p-islands on a silicon substrate. Besides the processor with its logic gates, the device contains other parts too. It includes some memory cells to store the data that the chip calls upon often (a program of instructions, for instance) and control circuitry that governs how the signals flow from one part of the device to another.

The processor and memory parts of modern computers, along with control-circuitry, encompass three of the five components that Babbage said computers must have. The other two are the input and output units. In Babbage's time these were mechanical devices by which the user put numbers into the machine (either data or instructions) and got other numbers out as results. Today, input and output units are still very important. The input unit of a computer could be a keyboard, a light pen (with which drawings are made on a screen) or possibly a microphone, if the computer has got enough clever circuitry inside it to understand human voices. The most common output device is a screen on which the computer flashes up lines of text or numbers. Other output devices are printers to give hard-copy prints of results, or rolls of tape or microfilm on which data are kept in condensed form. It is worth remembering that, however much microelectronics may reduce the size and cost of logic or memory components, it cannot do very much about the cost of input and output devices. These are subject to the cost curves of the non-semiconductor world which means, if anything, they increase in price. Cost reductions in microelectronics, however, more than compensate for any cost increases affecting computers from this direction.

Another thing worth remembering about modern computers is that not all their memory is likely to be in chip form (though their logic components will be 100 per cent chips). Most computers, especially fairly big ones, have back-up memory stores which record data on either discs or tape. These data are stored permanently (unless they are purposely erased) unlike the information stored in chips which only remains in place while an electric current is on.

The first microprocessor was announced in 1971, just twenty-three years after the birth of ENIAC, the pioneering electronic computer. Made by Intel of the US, the first microprocessor contained the equivalent of about 2000 transistors on a piece of silicon that could sit comfortably on a typewriter key. It was roughly as powerful as the thirty-ton ENIAC which, those few years previously, had given the world such little warning of the kind of machines that were to follow it.

The chips we have just mentioned are the type manufactured in their thousands coming off production lines in a similar fashion to any other mass-produced commodity, cars for instance. There are other, more specialised, kinds of chips suited to limited numbers of customers. Instead of being the building blocks of computers – and this means they eventually find their way into a breathtakingly wide range of applications – these chips will do equivalent jobs in other, more specialised, electronic systems. For example, they could be frequency converters in satellites, signal synthesisers in TV sets and radios, decoders or analogue-to-digital converters in industrial machinery, or special processor chips for telephone exchanges. British semiconductor companies, much smaller than their American rivals, specialise in these kinds of areas, leaving the giant US and Japanese firms the mass markets.

We need to stop a minute to think about the dimensions involved in the increasingly important engineering discipline of making microelectronic circuits. The crucial pockets of n- and p-materials are just a few micrometres apart, pretty small distances when you consider one micrometre is about the diameter of a speck of dust. Generally speaking, the closer these 'islands' are, the quicker the circuit reacts to an external signal. At the same time, the circuit will use less power because less energy is required to push an electron a short distance as opposed to a long one. Because of the tiny distances involved, it is possible to run microcomputers consisting of several chips off a power source no bigger than a torch battery.

Computer designers can make their circuits faster by injecting them with more current. As they don't want to put too much power into their devices (for one thing this manifests itself in heat that has to be channelled away somehow) they normally settle for a compromise between speed and power. The quickest logic gates in computers react to signals in a matter of 'nanoseconds', the unit of time that is an essential part of a computer designer's vocabulary. A nanosecond is one billionth of a second. To give you an idea how small a unit this is, light, which travels more quickly than anything else in the universe, manages about a foot in one nanosecond. By contrast, in one second, a unit of time we are more used to dealing with, light travels between here and the Moon.

Electronic engineers shrink their circuits to such tiny dimensions by what is called photolithography. This is the technique with which they 'draw' lines on a silicon substrate which mark the positions of the n- and p-materials required. First, of course, someone has to devise a circuit diagram that shows exactly where each n- and p-island has to go. Considering the vast numbers involved, this is by no means an easy job. Luckily, a computer helps. With the aid of a specially-programmed computer, the circuit diagram is drawn hundreds of times bigger than life-size. A computer can simulate the workings of another computer just as easily as it simulates, say, a tennis match or the operation of a spaceship in a TV game. In response to commands typed in by an operator on a keyboard, it works out in what order the circuit elements should be put to do a certain job, how many of them there should be, and so on.

The finished diagram, on a big sheet of paper, is checked for faults by humans (or by another computer program) and then stored in the computer's memory. A science-fiction writer could hardly imagine a better example of computers playing a key part in replicating others, and so ensuring the survival of the machine race.

The diagram, coded as a series of 0s and 1s, probably in a memory chip such as we have just described, is then transferred to the store of another computerised machine called a pattern-generator. This automatically defines the diagram, on the microscopic scale required, on to a piece of glass called a mask. The lines representing the positions of transistor-equivalents and so on are inscribed by electron beams that move back and forth in response to commands from the pattern generator. At the start of the process, the glass has a layer of chromium on top of it. On top of this there is another layer of resist, a chemical that is sensitive to electrons, and which changes in structure when they impinge upon it. The parts of the resist hit by electrons are dissolved away in developer solution, leaving the chromium exposed underneath. (The same resist principle is used in photography to make prints from film.)

The exposed chromium is then dissolved in some other solution that does not affect what is left of the resist layer, leaving tiny sections of the glass visible as a series of microscopic 'windows' surrounded by layers of resist and chromium. Finally the remainder of the resist is removed with another chemical, to leave the glass with a series of fine lines of chromium on top of it that mark out the design of the chip.

This is only the beginning of the chip fabrication story. In practice, the mask is made about 10 cm in diameter so it contains the designs for several hundred chips, all exactly the same. Then it is put in the path of a beam of ultra-violet light. Where this light hits the chromium, it creates a shadow. Where it meets the glass 'windows', it passes straight through to a piece of very pure silicon, the same size as the mask, called the wafer. This has a thin layer of silicon dioxide on top of it, with another resist, this time sensitive to light rays, impregnated on top of that.

In a similar fashion to the way in which the mask was made, the resist is developed and the exposed silicon dioxide removed to leave what is called a 'device mask', the silicon wafer overlaid with chunks of silicon dioxide that define a broken pattern. The pattern outlines a series of 'entrances' into the silicon through which the relevant dopant atoms are directed. This is done either with the diffusion process we have already encountered, or with another, newer, method called ion implantation which gives better control over the quantity and depth of the intruding foreign atoms. In this way a series of islands of n- or p-materials is inlaid into the silicon, defining part of the complete circuit pattern for the chip.

The engineer would then cover the wafer's surface with a fresh oxide layer and start the fabrication process all over again only this time with a different mask to impregnate more dopant atoms in different places. Perhaps six masks might be required to make all the dopant islands the chip needs, plus another set to define where the layers of metal should go to join the islands electrically. The final stage is when factory workers dissect the wafer into its constitutent chips, test them thoroughly (fifty per cent of the chips might fail the test and are simply thrown away) and put into packages that allow for electrical connections to the outside world. The packaging is what you see when someone shows you a chip: many people are a bit surprised at first because the 5-mm-square chip itself is largely hidden and the device looks much bigger than they have been led to expect.

Making chips, then, is an intricate business where little things matter a lot. Indeed, it could be said that never before in history has so much depended on a small bunch of people getting so many little things right. Apart from design costs (which we shall cover in the next chapter) semiconductor firms have to pay out hefty sums for the manufacturing equipment itself. An ultra-violet projection system costs about £100,000;

an ion implanter about £200,000 and a good electron beam machine £500,000 to £1,000,000. Because of the constant race to be at the forefront of new ideas and new designs, equipment is written off very fast. Texas Instruments, for instance, spent £25,000,000 on new equipment to make its new 64K RAM computer memory. This is apparatus that will be scrapped a few years after it has been commissioned because it is obsolete. Motorola, another big US chip firm, spent a similar amount tooling up for a new microprocessor, the MC68000.

A fierce race is now taking place to produce the next generation of machines that will use X-rays or direct-write electron beams (electrons which write directly on to a wafer without the intermediate step of mask manufacture) to make chips with the distances between dopant islands less than a micrometre. At present, most chips have circuit dimensions anything between two and six micrometres.

This chapter cannot close before we dwell a little on the kind of cost reductions the techniques just outlined have brought about. The position was put graphically in an article by Robert Noyce, one of the leading US semiconductor engineers, in *Scientific American* in 1977. It is a general rule in most industries, he said, that they reduce their costs (at constant prices) by between twenty and thirty per cent each time their output doubles. This is according to the theory of the 'learning curve' by which costs drop as experience increases. Noyce pointed out that the world's semiconductor industry has doubled its output, not once or twice, but eleven times in the seventeen years from the time it started in earnest in 1960. The number of individual components (transistors, logic gates or memory storage cells containing one bit of information) that the industry can sell in one year has soared from 500,000,000 in 1960 to a million million, or 2000 times that number, by 1977, leading to a hundredfold decline in the average costs for electronic components since the integrated circuit was born.

Even more staggering has been the cost reduction in one area, memory chips. The cost of storing one bit of information, that is, the cost of making one memory cell, has fallen by roughly thirty per cent every year from 1970 to the present. If the same price reductions occurred over the next decade in other, more familiar, areas of life, you can bet people would really sit up and take notice. By 1990 (at constant 1979 prices) a three-bedroomed house would cost £600, a new car £60, a gas cooker £5, a pair of shoes 50p, a gallon of petrol 4p, a packet of cornflakes a penny, while a local telephone call would cost one fifth of a penny. What a pity, you might think, that some bright engineer cannot think of a way of spreading the price reductions that have affected electronics a little further into other parts of our existence.

This hugely increased demand for electronic components, fuelled by the decline in costs we have just talked about, cannot be explained by any single market area taking up large amounts of semiconductors for specific use. Rather it has been the wide spread of applications for microelectronic parts that is behind the demand for them. This range of applications, marked in the past few years or so, will go on growing at an accelerating rate, as the rest of this book will make clear.

Part 2

APPLICATIONS

Chapter 4: Money talk, binary-style

If anyone queries why the world should get excited about the microchip age, send them round to the nearest financial analyst who has had his head screwed on properly. He will tell them, in precise business terms, that chips should be watched not because of any emotional nonsense about their effects on jobs or society as a whole, but because of the amount of money that is tied up in them. There is something in this hard financial view. Semiconductors now play a big role in determining how the world runs itself economically.

There are two ways of looking at chips from a financial standpoint – from the supply or the demand side of the economic equation. Let's look first at the demand for chips, largely created by the big cost reductions we have already talked about. The cut in costs of individual electronic functions such as logic gates and storage cells make the idea of computerisation – or the addition to fairly ordinary bits of equipment of 'computer intelligence' – more attractive to more and more people. Hence the drive by manufacturers to incorporate computer functions in their products, confident that the functions will increase the products' value and constitute a good selling point. This demand side of the economic equation is generating such a market for chip-based electronic products that by 1990 the electronics industry is likely to be one of the world's biggest businesses.

On the supply side, the chip companies, which we will talk about in detail later in this chapter, are geared up year-by-year to increase both the volume of output and to make their devices more sophisticated. So far there has been little problem finding uses for this growing annual output of chips: the world has lapped them up, apart from the odd spell such as around 1973 when global recession dampened demand. The industry has grown used, every year, to increasing output by fifty per cent and reducing the price of individual functions by thirty per cent. Whether it can continue to do this is difficult to answer. The technical advances that made it possible to cut prices with such regularity, thus fuelling the demand for chips, are becoming more difficult to make. Nonetheless the furious competition in the semiconductor world keeps the race on.

If one company fails to make its chips 'cleverer' or fails to cut their cost, then it can be pretty sure another firm will. The latter would thus gain a lead that could drive the first company from business. The technological battle requires money; lots of it. The semiconductor industry has an appetite for capital expenditure on new plant and processes, and a demand for the time and effort of skilled engineering staff, that is virtually unequalled in any other branch of commerce.

Let us look at the kind of semiconductors the world finds most suitable for its needs. Strictly speaking, semiconductors are divided into two categories – discrete devices and integrated circuits. (Thus semiconductors are not synonymous with integrated circuits, although many people think they are.) Discrete circuits are the 'old-fashioned' type of semiconductor in which there is just one electronic component, say a transistor, per sliver of semiconductor material. Integrated circuits contain more than one – as we have seen they can contain tens of thousands of components without any trouble on one chip of silicon substrate. Thus a discrete circuit becomes an integrated one when the lone component becomes joined by a few more of its fellows. The world bought about £5 billion worth of semiconductors in 1979, ten times more than in 1970. Of this total, two-thirds is integrated circuit sales – although this proportion will increase as the 1980s progress when the more densely-packed type of circuit, such as microprocessors and memories, become even more widely used than now.

Microprocessors and memories represent the very latest techniques in chip manufacturing. However it is worth remembering that in 1979 the microprocessor, the most talked-about type of chip, accounted for only seven per cent by sales of the world's integrated circuit output, and memories some twenty per cent. The rest of the chip market is taken up by non-programmable logic units for big computers' central processors, and other specialised chips such as analogue-to-digital converters and 'support' chips that have to cluster around a microprocessor for it to work effectively. For every one of the thirty million microprocessor chips sold worldwide in 1979 it is salutary to remember that perhaps a hundred much less densely packed logic chips went into the guts of the big mainframe or mini-computers. Microprocessors and memories will, however, form the core of the small, programmable, decision-making devices of the 1980s and the 1990s and by 1990 it is likely that they will account for two-thirds of all integrated circuit output. By this time, world production of integrated circuits will be around £15 billion annually.

Semiconductors account for an average of around five per cent of the price of a conventional electronic product, where the product could be a chip-aided toy, a TV, a communications terminal or a computer. You might think this percentage is not very much. Indeed it is not – the rest of the price is taken up by much more expensive items such as casings, wires and other bits of electrical circuitry, such as keyboards and cathode ray tubes. These were all around long before the silicon chip and have not dropped in price (probably just the reverse). Then there is the cost of putting all the bits together and making them work (this includes software costs), which is likely to be considerable.

Electronic products with the right kind of chips in them to make them programmable, or computerised, are called 'intelligent'. (Often the chips comprise one microprocessor plus support circuits.) This means, loosely speaking, that they are capable of making decisions. Thus an ordinary TV set is an 'unintelligent' – some people prefer the world 'dumb' – device. On the other hand, a computerised communications terminal, linked either by wire or radio waves to a computer, can not only receive and transmit signals but decide what to do with them according to how it is programmed. Thus it could print out messages on a printer or flash them on to a screen for visual display. Or it could route the signals to a machine like a lathe which would operate accordingly.

Intelligent machines generally need sensors that feed the computerised part of the device with information about what is going on in the outside world. Such sensors could be thermometers giving information about temperature, pressure gauges providing details about pressure, or some kind of radio or telephone apparatus that sends electrical signals about a particular quantity. The intelligent machine also requires devices called 'actuators' that try to alter the outside world in accordance with whatever the computerised, or control part of the machine, decides. Such actuators could be motors that drive, for example, a printer or a lathe that makes parts in a factory. Actuators and sensors normally account for a large proportion of the machine's cost.

The 'intelligent machine' can be represented as in figure 4.

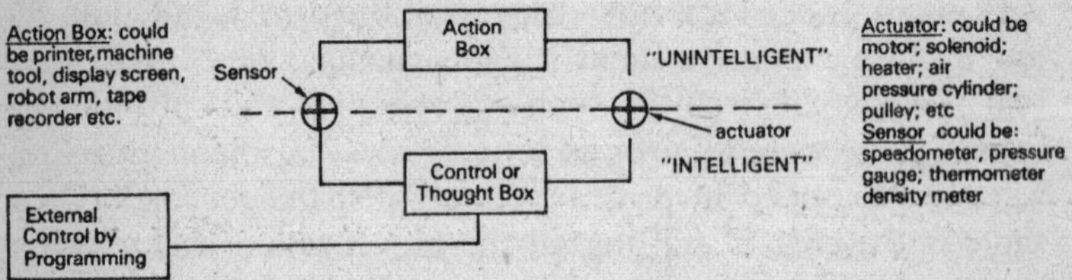

Fig. 4.

On the 'intelligent' side of the dotted demarcation line, the control box receives signals from the sensors, converts them to 0s and 1s of the computer code, sorts them out with the help of its memory cells and logic gates as shown in Chapter 3, and instructs the actuators accordingly. The piece of the machine on the other side of the line merely does what it is told.

'Thought' is on one side of the line; 'action' on the other. Thus the machine can be regarded in a similar way as a person whose brain sorts out how to tackle problems and whose limbs actually carry them out.

The advantage of having a computer on the control side of the demarcation line is that it can be programmed to act in different ways depending on its owner's whim; and it uses its intelligence to work out how best to do something from a large number of sensor inputs. This is by contrast to mechanical control devices, bereft of any intelligence whatsoever, which usually take into account just one variable at a time. For instance, an ordinary thermostat acts only upon temperature readings, and considers nothing else when ordering valves – the devices on the 'unintelligent' side of the line – to open or shut.

The control part of the equipment, even if it doesn't cost much and is so small that it can hardly be seen (you will probably need to prise off a good part of the device to be able to see the 'control box'), will be crucial to the product's chances of commercial success. This is where all the cleverness in the piece of equipment is lodged.

Semiconductors don't add many costs (not hardware costs anyway) but they add value. The extra value could be seen in, for example, a toy tank that a child can program in a rudimentary way to steer in a prescribed path, so giving it an advantage over rival weapons of war in the toddlers' club tank battle. Or the 'intelligence' could be evident in domestic appliances such as washing machines that take note of factors like temperature and humidity and 'decide' on the correct speed to run their motors, so saving energy. The extra cleverness might be apparent, on the other hand, in industrial equipment that controls chemical reactions better than non-chip-aided apparatus.

The total value of electronic equipment sold worldwide, a large amount of it containing significant quantities of chips, will come to about £90 billion in 1980. In rough terms, the US produces about half these goods by sales, Western Europe twenty per cent, and Japan and the rest of the world fifteen per cent each. Growing as it is by some ten per cent per year, the world output of electronics equipment could easily run to £180 billion by 1990. This will push it very high indeed in the world's commercial league table. According to the US consultants Booz, Allen and Hamilton, electronics will then be more important than the petroleum and electrical power generation industries, whose annual turnovers then will be about £140 billion. And the industry will rival car manufacturing, steel and chemicals which by 1990 will have projected annual sales in the £230 billion region. The size of these industries, the biggest that there are, can best be appreciated by comparing them to the UK's annual production of goods and services which is roughly £150 billion (in 1980).

What kind of goods come under the heading electronic equipment? It is these products, the application areas for chips, that we shall consider in detail in the later chapters. But to get a feel for the relative importance of the different types of chip-aided equipment, we can again borrow figures from Booz, Allen and Hamilton. In 1976, data processing (which means large computers and ancillary equipment such as printers, display screens and so on) accounted for twenty-six per cent of world sales in electronics goods. The US government was a big enough customer to warrant a category of its own, mainly covering purchases of defence equipment (see Chapter 9), and accounted for eighteen per cent. The next biggest classification was consumer goods with fifteen per cent; industrial equipment, such as control apparatus for factories, twelve per cent; communications equipment and TV were each responsible for eleven per cent; and office equipment, such as word processors and other computerised machines to help clerical workers, accounted for seven per cent. The fastest growing areas are in data processing, industrial equipment, consumer goods and communications apparatus where sales are increasing by sixteen, thirteen, twelve and twelve per cent respectively.

If we consider just the market for microprocessors in

electronic equipment, then, according to the consultants Creative Strategies International, the terminal/computer market will see the biggest growth over the five years 1978–83, increasing from sales of £36 million annually to £100 million. The consumer market for microprocessors will experience the next biggest surge, from annual sales of £30 million to £80 million followed by transportation which will consume £50 million worth of microprocessors in 1983 compared to £10 million in 1978, the increase being largely due to new demands for microprocessors in cars. The consultants say the total demand for all microprocessors will rise from £160 million to £450 million over the five years.

Some people might say the future is one of uninterrupted boom for chip firms as, one by one, every conceivable piece of equipment in the world has a microprocessor or other assembly of chips incorporated in it to raise its performance to unheard-of heights. Such a scenario, however, appears unlikely to all but the most fanatically enthusiastic proponents of chips. Some semiconductor pundits have gone so far the other way as to prophesy the day when the chip world faces a problem of demand saturation, when there are simply not enough uses for all the chips that the world produces.

Such a situation is not all that difficult to imagine – when you consider the huge quantities of semiconductors that the world produces. Gordon Moore, president of Intel, one of the world's biggest semiconductor firms, summed up the dangers in a speech in 1977. Assuming that chip production would go on increasing by something like fifty per cent per year, Moore said that at its current rate of progress, the industry could produce a hundred thousand billion (10^{14}) circuit functions in 1985, where a circuit function is either one logic gate or one memory cell. (The functions would of course be integrated in 'job lots' of several thousand at a time, on to chips.) This huge total works out at some 50,000 functions (which on Moore's reckoning is roughly equivalent to one mainframe computer) for every two people on Earth. Moore's message, a frightening one to the semiconductor industry, was that total output in 1985 would be so big that nine-tenths of it would simply stay on the shelf unused unless the business made an effort to develop new applications which could soak some of it up.

Moore's apocalyptic words bring into focus the perennial problem facing the semiconductor industry. Firms must on the one hand decrease the costs of their devices year by year to keep up with the competition. On the other hand, they must finance the huge research and development that the cost reduction entails by somehow selling the cut-price devices to an ever-growing network of customers, assuming that they can be found. In any other business, this would be a recipe for suicide. That the chip industry is special is proven by looking at its incredibly successful record so far. Yet even allowing for the huge number of uses to which computerisation or 'added intelligence' can be put, there is the obvious danger that prices could fall so low that the industry simply cannot find enough customers to spread its costs over. In other words it could be forced to a point where the millions of extra chips it makes to recoup costs remain unsold.

The invention in 1971 of the microprocessor came just in time to solve this problem at least for a while. Hitherto, the semiconductor industry made and sold a lot of individual devices that did not have that much of a wide appeal. They were attractive only to those customers who could use them directly for particular jobs. The microprocessor, however, started the trend of chips that could be programmed – just like the big old-fashioned computers of the 1950s and 1960s. Chip firms can sell microprocessors to different types of customer who adapt the devices to their own needs simply by changing the software. In this way, the big development costs of one brand of microprocessor are spread over a huge number of customers. What the chip industry could do with to solve all its problems is the invention of something like the microprocessor every five years or so to give it a new lease of life, but such an eventuality is almost impossible to imagine.

Perhaps a more likely scenario is to find, every few years, new customer groups to which large numbers of existing products, for example microprocessors, could be sold. This in effect is what has happened over the last few years of the 1970s as chip firms have latched on to the consumer market. This is where the action is going to be in the 1980s, according to many in the semiconductor world. Although devices such as chip-games (for instance the ones that plug into TV sets) and gadgets for the home, like microprocessor-controlled washing machines or fridges, are difficult to plan for because the direction of sales may change altogether as a result of changes in taste, the market is extremely responsive to advertising pressures – a big 'plus' for the chip firms. The gadget-loving American household is the initial target for the chip firms, or more directly the people who buy semiconductors and put them in their own devices (original equipment manufacturers or OEMs in the industry jargon). To the question 'Why do you make chip-enhanced novelties for the home?' the chip marketing man would answer very simply: 'Because you can sell them.'

With fairly straightforward arithmetic, it is possible to demonstrate just how big the consumer market could be for the chip industry. In the peculiar numbers game that semiconductor people play when calculating who will buy their products, Gordon Moore worked out that if all fifty million American households could somehow be persuaded to buy the equivalent of a million electronic functions per year in gadgets for the home, then his projected 1985 shortfall in demand of 9×10^{13} functions could start to be eaten up. A target of a million functions per home is really not all that much. For example an enhanced version of Texas Instruments' Speak and Spell learning toy, a 'talking calculator' educational aid that sold for $55 in 1979, contains about 800,000 memory cells with just a couple of extra 256K RAM modules added to the basic model. A device like this sold to half America's households every year would at a stroke account for 4×10^{13} electronic functions, or half of Moore's supposed shortfall.

The point has not been lost on the world's chip makers. For instance, Bill Howard, director of technology and planning at Motorola, says that the consumer is a long way from appreciating the full impact of microprocessors: this will come during the 1980s as they enter the home at a great rate. Howard does not think people will buy gadgets just because they have microchips in them; just the reverse. He says the vanishing computer will make its entrance accompanying everyday objects, with people noticing only the enhanced control or added feature, such as saving energy in heating appliances (or saving soapsuds in a micro-controlled washing machine) and not the chips themselves. 'If the consumer is made aware that a computer is in a device then this will be a block against selling it. As excited as people get about developments in microelectronics, the rate of change in society is slow. Therefore the use of microelectronics will be in those things that exist already.' We will look further at chips and the world of consumer gadgetry in Chapter 6.

We have discussed something of chip firms and how they see the future. But exactly who are they? Texas Instruments, Motorola and Intel, who have been introduced already, are the world's top-selling semiconductor firms, in that order. They are from the US, like most things to do with microelectronics. In 1979 the US accounted for sixty per cent of the world's semiconductor production, and ninety per cent of microprocessor sales. After the top three come a clutch of other US firms such as National Semiconductor, Fairchild, Advanced Micro Devices and Mostek. Japanese firms are well down the list, although they are gaining ground rapidly (see Table 1).

Many of the top US firms are based in the area known as Silicon Valley just south of San Francisco in California. This is where the chip industry sprang up in the 1960s after the invention of the integrated circuit in 1959. The roots for the industry grew from the most successful of the early semiconductor firms, Fairchild, which a group of electronic engineers founded in 1957.

TABLE 1: Annual sales of top US and competing countries integrated circuits firms. Source: *Business Week* and the Economist Intelligence Unit (1979 figures are estimates)

	Integrated circuit sales ($ million)
	1979
Texas Instruments	780
Motorola	450
Intel	430
National Semiconductor	420
Fairchild	330
Signetics	250
Advanced Micro Devices	210
Mostek	200
RCA	145
Harris	120
	1977
Nippon Electric	156
Hitachi	114
Toshiba	98
Fujitsu	72
Matsushita	53
Mitsubishi	48
AEG Telefunken	all 15–20
GEC	
Plessey	
Ferranti	
Thomson CSF	

Thereafter many of Fairchild's employees left to establish their own firms in or around Silicon Valley. There were so many of them that they came to be bracketed as 'the Fairchildren'.

Many of the chip firms started up from virtually nothing in the early 1960s when the 'entry stakes' – the costs of research, capital equipment and so on – were relatively low. In recent years, however, the amount of cash needed to finance developments has grown larger as technical problems have increased. Banker Ming Li of the First Boston Corporation, New York, has estimated that in the early 1960s, when the semiconductor industry was in its infancy, it needed to spend thirty cents to generate sales of one dollar: in 1979 the figure was seventy cents, more than twice as much.

Much of the cash must be spent on research and development. Semiconductor firms typically spend ten per cent of revenue on this, a much larger proportion than the rest of industry where the figure is nearer two or three per cent. The research effort is in two main areas. First engineers must continually develop new techniques of physics which will squeeze more components on to the same 5 mm square piece of silicon. As the separation between the components becomes so incredibly small – the standard chips of 1979 have component separation of about four micrometres and this will be down to two micrometres by about 1983 – the technical problems become immense. Precision engineering at these tiny distances – in the world of the so-called 'micro-universe' – requires skills that today's technologists are still learning about.

The second area of research is in trying to design chips' layout better so that, in a new chip, the same number of components as in an earlier model do a better job, so making it easier to use and cutting running costs. This activity concerns component 'architecture': designers have to work out in which order to put their building blocks of gates, memory cells and so on, in the same kind of way as architects have to put together things like doors, windows and bricks when constructing a house. Much of the costs of improving chip architecture go towards designing better software that will run on a certain semiconductor layout. As software costs are mainly the time it takes for highly-paid computer programmers or designers to work out a series of highly complicated problems – and this means a long, long time – software expenses take up a high proportion of most firms' research efforts.

Motorola spent $10 million designing its new microprocessor called the MC68000; Intel twice as much on its broadly equivalent 8086. Of these sums, the software design probably takes up eighty or ninety per cent. The huge amounts of cash needed to be a force in microelectronics explains why the day of the small chip firm is almost over. Many of today's semiconductor companies are at the very top of the commercial tree. Texas Instruments and Motorola, the two companies that sell the most chips, have many other interests besides integrated circuits including car radios, computers, home entertainment systems, and various types of computerised communications equipment. They have annual turnovers of $2.5 billion and $2.2 billion respectively (1978 figures) when all their business interests are totted up. The two firms' capital expenditure in 1978 amounted to $300 million and $146 million, more than most British firms' annual sales. In the 1970s, only three semiconductor firms started up from scratch, Synertech and Zilog (owned by the huge oil firm Exxon) and Britain's Inmos. Yet there has been a plethora of mergers and joint efforts between companies which have semiconductor interests, indicating that the cash needed to keep in the semiconductor race is just too great for medium-sized companies to go it alone. Examples of such events are the take-overs by two giant firms, United Technologies and Schlumberger, of Mostek and Fairchild; the purchase by the Netherlands electrical company, Philips, of the US's Signetics; and the proposed joint venture between Britain's GEC and Fairchild (Schlumberger, Fairchild's new parent, backed out of the deal in 1980.)

When considering chip firms, we must not forget the world's biggest computer company, IBM. With annual sales around $17 billion, IBM is responsible for sixty per cent of the world sales of computers. Although it does not sell chips, the company spends a lot of effort making them for its own products. For example it was one of the first firms to announce it had produced the new generation memory chip, the 64K RAM, even though it did not offer it for sale to anyone. It is estimated that IBM spends $300 million yearly on integrated circuit research. Including neither IBM nor Bell Laboratories, another large company that makes chips for its own use in telecommunications equipment, the US chip industry spent about $200 million on research in 1977 alone.

So much for the US industry, which is where integrated circuits began. What about the rest of the world? The Japanese are without doubt the next strongest in the semiconductor league table. They are catching up so fast on the Americans that many US chip firms are worried they will be overtaken – both in technological know-how and in expertise in marketing semiconductor products. Japan accounts for some thirty per cent of world semiconductor sales. Its strong position has a lot to do with government decisions that a big technological effort in this area could do much to boost the country's export industry in electronics. As long ago as 1971 Japan's Ministry of International Trade and Industry decided to back semiconductors with a series of impressive research and development programmes. In one of them, to find new methods of large-scale integration, the government spent £50 million over the last few years of the 1970s. So successful has it been that Japanese chip firms are predicting they will have chips with a million components on them by the early 1980s.

Governments in Europe realised rather later than the Japanese the importance of microelectronics. (The US government, through its Department of Defense, whose involvement in semiconductor research programmes goes back to the Vietnam War, saw the light even earlier.) According to estimates by the British government, both France and West Germany are supporting microelectronics development programmes in their country's industries to the tune of about £100 million each, in both cases the money being spread over five years.

The British government seemed to wake up to microchips only in 1978, although the UK Ministry of Defence, like its American counterpart, had supported chip developments for defence equipment for many years before. In a series of moves in 1978, the then Labour government poured finance into several ventures either directly or indirectly concerned with microelectronics. The government promised to spend £55 million educating industry in general about microprocessors after a survey revealed only five per cent of UK industries were using them in their products or production processes. Through the National Enterprise Board, the government agreed to spend £50 million and £40 million respectively in helping to set up brand-new firms, Inmos and Nexos, which would make new generations of microchips and semiconductor-aided office systems. The Prime Minister James Callaghan spelt out what his government thought about the issue when he said in December 1978: 'Microelectronics is of crucial importance to the industrial future of the nation.' When the Conservative administration took power in May 1979, it did little to change the emphasis the previous government had put on the importance of the new technology, apart, that is, from expressing a desire that developments in microelectronics would be best carried out under the wing of private firms. Hence Industry Secretary Sir Keith Joseph asked the National Enterprise Board to sell off its holdings in high-technology companies such as Inmos as soon as these firms are strong enough to attract private purchasers.

A question that many people ask when told about developments in microelectronics is: 'Why do they happen?' In other words, they question the need to improve chips year by year so that each new product is more sophisticated than its predecessor. Is there no natural limit, they ask, for example when semiconductors are developed that can do the job of the human brain, at which point the whole process will stop? Most semiconductor people would say there is not. What drives them on is money – money for themselves and for their firms if they can somehow continue to make devices that are better than their rivals' – and, we can say rather more charitably, the pursuit of engineering excellence. Intel, the chip company that started from nothing in 1969 and is within sight of becoming a billion-dollar turnover firm in the early 1980s, symbolises the attitude of the industry. Gordon Moore, Intel's president, summed up the position at a meeting in April 1979: 'Intel is dedicated to keeping the (cost reduction) machine running. We are not near any fundamental limits. There is plenty of room to make at least another ten or hundredfold cost decrease.' With this thought in mind, that chips will continue to become more sophisticated, more powerful and less costly, and that these developments will have a big effect on the kind of cheap intelligent machines that appear in the future, let us go on to consider in more detail those areas in which chips will make their presence felt.

Chapter 5: The great communications explosion

Around the year 2018, if present trends continue, an event will occur of startling significance for the way we run our lives. The event concerns the transfer of information, and how this is carried out over long distances. When you think about it, information transfer is one of the most fundamental activities there is. Seeing, feeling, speaking, hearing – all are names for communicating information from our surroundings to our brains so that we can enjoy life, talk to people, do business and so on. A great way of transferring information over long distances is by the telephone system. This is done simply by picking up a telephone handset, dialling a number and exchanging information by talking to whoever is at the other end.

So what has the year 2018 got to do with all this? It is simply that in that year, if things carry on as they are, the nature of the information communication game will be changed irrevocably. Up until then there will always be more telephones in the world than computer terminals, which are a relatively new way of exchanging information over the telephone system but using a computer to do it. In 2018 there will be, for the first time, more computer terminals than telephones. Communicating over long distances with computers will have become more important than performing the equivalent function with people.

You will ask how I have worked this out. The conclusion follows from an admittedly rather speculative piece of mathematics. The world in 1980 contained roughly 400 million telephones and a mere four million computer terminals. Telephones are multiplying at six per cent a year while computer terminals are surging ahead at something like twenty per cent per year. If these rates continue, by 2017 there will be roughly 3.4 billion of both kinds of device – thereafter the computer terminal will start to pull ahead.

The detailed accuracy of this forecast is of course open to question. The instability of the world's economic situation could play havoc with how both types of device increase, and change my figures accordingly. But this kind of argument is for the nit-pickers. What is certain is that communicating with computers is already widespread, and is becoming of even greater importance at a furious rate.

You communicate with a computer when you feed information into, or extract it from, a computer over the telephone system. (Here the word computer can be interchanged with the term 'computerised device' which means anything controlled by a computer, such as a microprocessor-controlled machine tool or an 'intelligent' or programmable terminal). Essential is a computer terminal, which sends out electrical signals along the wires and radio links that constitute the telephone system to the distant device. The electrical signals are coded so that they mean something in the 0s and 1s language of computer 'talk'. A form of computer communication even more simple to grasp is created when two bits of computerised equipment are connected so that they chat away to each other in their very own language of rapidly-changing electrical pulses.

We shall look at just why people want to exchange information with a computer (some may think this an odd activity) and the different forms of computer–terminal link later in the chapter. But first let us examine a few figures that demonstrate the huge amounts of data (what scientists call computerised information) that will be passed along the world's telephone system in 1990. At the rate computer information traffic is increasing, by 1990 there will be something in the region of thirty million computer terminals scattered around the world. Between them they will transmit the equivalent of 2.8 billion pages of typed script every day. It takes an effort to imagine what this represents. If this daily data dose were put in book form, it would just about fill eight million volumes, each the length of the one you are now reading. Left on ordinary typed sheets laid out end to end, the information would form a vast strip running for half a million miles, or twenty times the Earth's circumference.

Put yet another way, if the typed pages were stacked neatly on top of each other in time with how the information store is growing, the pile would climb to the top of St Paul's Cathedral in a little over thirty seconds. It would reach the summit of Mt Everest in around half-an-hour; and after twelve hours, the pile would be 100 miles high, far enough from the Earth to be immune from gravity, whereupon the huge quantity of paper, steadily becoming even bigger, would simply float away into the outer reaches of space.

Just what the world would do with this great volume of information if it didn't have cheap computers, and a reasonably efficient telephone system, to handle it, is an interesting point. Would people find they could get along without the information after all? Or have cheap computers entered the scene just in time to save us from drowning in a sea of paper (and using up the world's trees at a tremendous rate)? You can perhaps mull on these possibilities as we consider the mechanism that makes computerised information traffic possible – the ordinary telephone network.

Sir Edward Fennessey, a former deputy chairman of the British Post Office, in a statement of justified hyperbole has called the global telephone system 'the most complex machine yet constructed by man'. Representing a total investment of some £150 billion, the system affects an enormous number of areas of life, and the telephone handset – what you attach to the network to make use of it – has been one of the most popular scientific inventions, measured in terms of use, of the past century. In 1920 there were twenty-five million handsets in the world; in 1940 there were forty-five million; in 1960, one hundred and thirty million; and today, some three times that number.

Yet for all the activity and growth in the world system over the past half-century, the future, holding out the prospect of widespread expansion to the network of computerised devices as well as handsets, looks even more exciting.

The telephone, as used conventionally as an aid to long-distance conversation, carries electrical signals representing people's speech. Effectively it lets people transfer information to each other. Thus the telephone, when carrying voice traffic, can be termed an information system – of the same genre as a computer information network. However, it is not a very efficient one.

To say this is not to depreciate the powers of humans, nor to underestimate the importance of the telephone system for relaying the sounds of people's voices. But we cannot escape the fact that people communicate in a very inefficient manner. Their talk is punctuated by expressions such as 'you know' or 'how are you?' and all manner of jokes, anecdotes and remarks about the weather which may make talking to a person a pleasant business – but interfere dreadfully with the nitty-gritty of transferring information.

With computers, the situation could not be more different. Their language, composed entirely of 0s and 1s, each digit meaning something very definite, precludes the kind of redundancies that occur in speech. Thus computers are very efficient at transferring information – and they do it very quickly too, at many thousands of binary digits per second.

The kinds of people who get involved with communicating with computers are earmarked in a 1974 study by PA Management Consultants of London. According to the report, manufacturing industry accounted for twenty-nine per cent of data traffic passed along telephone lines in 1972. Data processing services (companies that carry out computer jobs on their own machines for customers who may be hundreds of miles away) were responsible for twenty-two per cent; banks for fourteen per cent and scientific establishments for twelve per cent. (The study predicts these percentages to change to twenty, forty-four, nine and eight respectively by 1985.) The remaining users of computer communications services are split between a wide range of organisations such as news agencies, airlines and government departments.

Let us look at the kind of jobs these organisations use their computer links for. Shops and offices use computer connections to update stock figures and match them with others at distant locations. The world's airlines have their own computer network that arranges aircraft reservations in more than one hundred countries. This handles every year some fifteen billion words of messages, each put in the form of binary code and transmitted as electrical signals sent along the telephone system.

Five hundred of the world's banks also have their own private network that every day sends about 30,000 financial messages from computers or terminals to others scattered across the globe. There is a world meteorological network in which computers in different countries exchange information on the minutiae of weather patterns, a far cry from the days before computers when met. men could do little more to advance their knowledge of the world's weather than ring observers and ask if it is raining. Scientists are big computer communicators. They 'dial up' computers, often hundreds of miles from their laboratory experiments, and ask them nicely – in computer language of course – to analyse their results. Personnel managers who want to work out staff salaries, local authority officers calculating rent arrears, accountants doing payroll checks – such people will similarly dial distant computers to do their jobs for them. Or if these people don't themselves own machines, they will 'rent' time on others run by computer bureaux.

A device that is becoming increasingly important is the word processor. It is useful in computer communication because we can think of it both as a terminal and as a computer – that is, it is an intelligent terminal. Word processors are typewriters that come linked to a computer, normally a microprocessor, with an added extra being the ability to communicate with other similar machines along telephone wires. One advantage it has over the orthodox typewriter is that the operator can 'call up' information from the microprocessor's memory and instruct the machine to print it. In this way the word processor can churn out thousands of copies of standard business letters, with perhaps just the name of the person to whom the letter is addressed changed on each one.

A second big advantage is that anything the typist writes remains in the computer memory until he or she tells the machine to print it. He can erase or alter words, letters or even whole sentences by instructing the memory accordingly – with the results of what he writes being flashed up as lines of text on a screen like a TV. Only when the typist has got his message word-perfect does he press a button and commit the whole lot to paper. Because there are no crossings out, or fiddling with whiting-out fluid, the typist saves a considerable amount of time.

Word processors today cost between £3000 and £15,000 depending on the sophistication of the machine. According to a report on word processing by the trade union APEX, in 1979 there were about 7000 machines in Britain compared to 350,000 to 400,000 in the US and 45,000 in West Germany. Sales in most countries throughout the 1980s are predicted to rise by thirty per cent per year.

We shall look in Chapter 7 at ways in which word processors could greatly affect the way in which office workers such as secretaries do their jobs. But word processors could usher in a more general change in the way that they make feasible what is known as electronic mail. This is where the word processor's ability to communicate becomes important.

Electronic mail is defined as any way of transmitting written messages using computerised signals passed along the telephone network. Under this broad heading, almost any form of computer communication could be described as electronic mail, although it is normally reserved for messages that would otherwise be delivered in letter form. According to the US corporation Xerox, more than seventy billion pages of information are distributed annually just among business users in North America. At the moment, less than one page in a hundred is transmitted electronically – the rest go by the traditional routes as letters or memos. But this percentage is likely to become a lot higher in the late 1980s, thanks to computerised devices such as word processors and another important kind of electronic mail terminal, the facsimile transmission machine.

Such so-called 'fax' machines are nothing new in concept. Fairly rudimentary versions have been around for years. They 'read' documents, rather like a photocopier, and then, instead of duplicating the page, they send signals representing what is on it along a telephone line to a receiving machine. The earlier versions work in analogue signals, but in the new sophisticated ones the signals are binary. This makes the machines much quicker, and, as the price of sending a facsimile of a document depends on how much it costs to open up a telephone line for the time the machine operates, cheaper too. There are about 500,000 fax machines in the world – 40,000 in Europe and the rest mainly in the US and Japan.

Most developments in electronic mail will affect the business user before the ordinary consumer. Xerox and two other American companies, AT and T and Satellite Business Systems (partly owned by computer giant IBM), are each planning their own electronic mail systems, with satellite links an important part and aimed mainly at big business customers, which will start up by the mid-1980s. The US Postal Service, however, wants to be allowed to start an electronic mail system which could affect the home. In 1979 it started tests on fax devices which read at ten pages per second, some 500 times faster than standard analogue fax machines, in preparation for such a service.

For most people, the nearest thing to an electronic mail service at present is the telex system. In this, customers type their messages into a machine which converts each character in the text into a characteristic pattern of holes punched in paper tape. When the typist has finished, the machine converts the patterns into electrical signals which are sent along the telephone system to a receiving machine, where they are reconstituted into print. Telex machines use the same network as the telephone, but they have their own special exchanges. There were about 100,000 telex machines in the world in 1979. If you don't think this is all that many, remember the German technician who in the 1930s said the world saturation level would be 3000 machines – his contemporaries scorned him and said he was a dreamer. This is yet another example of people failing dismally to anticipate the future. In today's computerised world, many of the existing telex machines will eventually be replaced by terminals similar, if not identical, to word processors that convert characters instantly to binary signals (without the intermediary of the paper tape) and are much faster to operate.

The crucial factor that could stop the development of a full electronic mail service is, fairly obviously, that not enough people have terminals. One way of surmounting this is to equip ordinary post offices in big towns with terminals, and persuade people to use the offices as electronic message centres for relaying information along the wires between them. In the local area around the post offices, postmen would continue to distribute the messages in conventional letter form. The obvious saving is time. Theoretically at least, the time taken to send an 'electronic letter' to someone hundreds of miles away depends on how long it takes to reach the local message centre, added to the time taken to deliver it from the centre at the other end.

This concept has certainly proved popular in the US where it is the basis of Western Union's Mailgram service, which started in 1970. In the scheme, 140 post offices in key areas have terminals and act as message centres similar to the ones just described. Between them they handled some thirty-three million messages in 1978, up from twenty-five million in 1976 and twenty million in 1975. The logical step from here is not that hard to visualise. Homes and offices could each have their own Mailgram-type terminal, costing perhaps not much more than a few hundred pounds, to send and receive messages without the 'middle man' of the message centre. The US Postal Service says that by 1990 such a system could, if it gets going quickly enough, handle half the seventy-five billion letters sent through the United States postal system every year.

The US is not the only country interested in electronic mail although it is far and away the most active. West Germany plans what it calls a teletex service of interlinked word processors which could within a few years link half-a-million users (mainly businesses) throughout the country. The UK, France, Japan and Switzerland have all experimented or are thinking about experimenting with public facsimile transmission services which – one day – could ensure almost instantaneous document delivery from one home or office to almost any other in the country, or for that matter the world. If this day dawned, the job of the postman employed to walk around dropping pieces of paper through people's letterboxes would be as outmoded an occupation as, say, a stagecoach driver in an age of cars and railways. In the postman's place would be an intricate system of wires and satellite links, connected to a multitude of little microprocessor-controlled display screens (acting both as computers and as terminals) flashing out messages every minute of the day.

The word processor, or any kind of intelligent message terminal which is both a computer and a message sending/receiving device, will loom large in our lives. It is communication between such devices that we will quickly grow used to. But there is another type of computer information network that will become almost as important. This is communication between a home terminal and a far-away computer – a big one this time, not just a little microprocessor – that is crammed full of facts, some of which we want to find out. The principle is simple. Take a quantity of information of the sort normally found, say, in books or scientific papers, and store it in a computer. Then allow other people to consult your computer 'data base' by dialling into it over the telephone system, charging them for doing so of course.

In 1979, this was big business for people interested in commercial or scientific information. There were about 500 publicly accessible data bases of this sort, containing some sixty million records. What happens is that companies store in their computers as many as fifty different data bases and work out an 'indexing system' that lets the customer 'scan' through the computer over the telephone to find the facts he wants. These facts are normally references to articles contained in libraries that the searcher then has to look up, the idea being that the computerised part of the process saves him the trouble of ploughing through a manual reference or indexing system. In due time, however, computers will probably store all the information the person needs, not just the reference, which will then be available at a touch of a few buttons on his terminal. The two biggest companies offering data base services over the telephone are both American. They are Lockheed, which nets about £12 million annually in this way, and Systems Development Corporation.

A relatively new concept is the consumer-oriented data base. Such data bases are now slowly starting to appear. Among them is Prestel, the British Post Office's Viewdata system. They operate in much the same way as the commercial data bases, but they contain less esoteric information – for instance on anything between where to go out for dinner at night to details of gardening clubs or railway timetables. Such Viewdata systems use the ordinary TV set, modified electronically, as an unsophisticated, but reasonably effective, computer terminal. The user controls it with buttons on a small key pad and the information, carried by the telephone system from the computer store to the home terminal, flashes up as lines of text on the screen.

Alex Reid, the PO's director of business systems, sees Prestel's role as similar to that of a magazine publisher in offering the public information that it would otherwise get from sources such as books and newspapers that are becoming increasingly expensive. The obvious parallel is with TV, that can be termed a passive information system, since you only have to turn one button to get anything from it. This has taken the world by storm over the past thirty years. According to Reid, and others backing viewdata projects, there is every reason to think this new system of active information transfer (you have to go out of your way to get it) will do likewise.

The immediate problem for Prestel is its cost. In 1979 the price of renting a viewdata set was £20–£25 per month, some £10–£15 more than it usually costs to rent colour TV. Also Prestel sets were in short supply. Both factors meant there were only about 6000 users of Prestel in Britain by the autumn of 1980. Reid sees costs decreasing to the point where there are half-a-million Prestel sets in homes and offices by the mid-1980s. However, given the numbers now using Prestel, this figure looks absurdly optimistic.

Other countries besides Britain are interested in viewdata. The telecommunications authorities (that is, bodies equivalent to Britain's Post Office) in France, West Germany, Japan and Canada, and several private companies in the US plan services similar to Prestel. To add to the activity, several countries are investigating or experimenting in teletext systems, where the principle is similar to Prestel but where the information is sent out as radio waves to modified TVs, not along telephone cables. Teletext services in Britain are Ceefax operated by the BBC and Oracle run by the IBA.

Electronic gadgetry is being fitted not only at terminus points on the world's telephone system. Also aiding the process of information transfer are computerised components that are starting to appear in the exchanges, or switching centres, that route messages from their starting point to their destination.

To help understand how these components do their job, it is worth backtracking on how we make ordinary telephone calls. The telephone system is in three parts: the handsets with which people talk and listen to each other; the cables and radio or satellite links that carry the electrically coded messages; and the hierarchy of exchanges that route calls. The need for the exchanges is fairly obvious. If there weren't any, each telephone in the world would have to be linked by a separate cable to all the 400 million others. The tangle of wires that would result is mind-boggling.

The coding in the system is an electrical signal that changes in accordance with the rise and fall in tone, and the stops and starts, of the human voice. It is a constantly varying signal that changes in time with the changes in the quantity being represented – thus it is an analogue signal.

Computers, as we know already, don't understand analogue signals. For them, the only things in life that count are binary signals, blips of current that say either 'on' or 'off'. The fact that the telephone system as it is constituted doesn't understand binary signals – why should it when for years it has had to deal only with analogue fluctuations representing the human voice? – poses obvious problems for computer communication. What happens when you send a series of blips along the telephone wires is that all is well until it reaches an exchange. And there the signals meet with the electronic equivalent of a blank stare and do not get through.

Thus to use computerised equipment in the ordinary network, engineers have to insert special devices called 'modems' that turn binary signals into analogue ones in between the device and the telephone system. They put a similar device at the receiving end to turn the analogue signal back into a binary one that the computer or terminal can understand.

This idea works very well, up to a point. It is of course messy, all this business of changing backwards and forwards from analogue to binary and back again, and engineers, most of them purists at heart, can't help thinking it would be nice if it could be eliminated. A more important problem is that the use of 'modems' limits drastically the speed at which computers can transfer their information. Computers can understand data transmitted at hundreds of thousands of binary digits per second. But the telephone network as constituted for analogue signals allows them to transmit it at the analogue equivalent of no more than 10,000 or so bits per second at the very most. Putting a computer in this situation is like asking someone to race a Ferrari along a cart track.

There are two main ways of overcoming the 'speed limit' by keeping the transmission all-binary with no break to analogue. The first is to lease private telephone lines from the relevant administration (in Britain the Post Office) rather than use the public network. The private lines can link the two computerised devices that the operator wants to take part in the information transfer. Thus there are no exchanges for the 0s and 1s of binary coded information to fight their way through: additionally, the lines can be made to a high quality to help achieve a high transmission speed. The disadvantage, however, is that the equipment's operator cannot place computer calls with whatever receiving device he chooses, as he can on the ordinary switched network, but has to be content with the same pair of permanently-connected devices.

In the second, more radical, method of speeding things up, this disadvantage is overcome. The exchanges that understand analogue signals are replaced by ones that deal with binary codes instead. Such exchanges can be thought of as yet another type of computerised device that receives a set of binary signals from one computer perhaps hundreds of miles away, decides on the basis of a code attached to the signals, where they are to be sent, and dispatches them along the correct wire to the receiving device. A few years ago, people would have laughed at the idea that anyone would spend time and money building a telephone network specially for computers. Yet this is what is happening in most parts of Western Europe. West Germany and the Scandinavian countries already have their purpose-built computer networks in operation, featuring both binary transmission and binary routing through computer-like exchanges. Britain started its own computer network, called the public packet-switched system, during 1980. A Pan-European network called Euronet run on similar lines to the national systems made a limited start in 1979. In the United States, where there are no government-controlled telecommunication administrations as occur in Europe, several private companies are either planning or implementing all-binary networks.

Where do the rapid strides in providing the means for computers to talk to each other leave good old-fashioned voice communication? Computer buffs might think this idea is out of date but it is a long way from being dead. Indeed the importance of voice communication is growing year by year. It is estimated there will be 1000 billion personal telephone calls made in the year 2000, up from 500 billion now and 250 billion in 1965. So there has been plenty of impetus for engineers to study how to improve matters for users of the voice network, as well as for those interested in data traffic. There is an old adage – if you can't beat them, join them. And this is exactly what has happened to the backers of the voice network. Most of the developed world's nations are now in the process of changing over their analogue-based telephone systems to ones that recognise binary signals and include computers in exchanges. Where once computers had to be adapted to meet the requirements of an analogue network meant to carry voices, people are having to adapt themselves to fit in with a world network that will eventually be totally computerised. Nothing signifies better the computer take-over.

Let there be no mistake: the computerised network will do the job of handling voice traffic better than the old analogue system. There are two important reasons for this. Binary signals passed down wires are less disrupted by electrical interference than analogue ones and so are easier to transmit. A 'rogue' electrical signal may jumble up the complex shape of an analogue wave form so it is unrecognisable and means nothing intelligible. But if the same thing happens to a binary signal – which is nothing more than a stream of pulses of the same height separated by gaps – it will usually be possible still to make out the bit stream's basic shape even if the tops of the pulses are whittled down a little. Thus it will come about that by the end of the century virtually all conversations carried by the UK's telephone system will be by courtesy of computer code. People will not have to know, let alone speak, binary language of course. Electronic circuitry built into the telephone handsets of the future will carry out the translation of voice signals into blips of current that are either 'on' or 'off', totally automatically. The on/off pulses will not change in format – only a computerised eavesdropper would be able to tap the line, not someone equipped with old-styled analogue equipment – until they reach the person at the other end of the line. Then they become reconstituted as sound waves that people's ears make sense of. In the long run, of course, the computer revolution may go so far that the binary boundary is pushed back still further: people may actually be able to understand computer 'blips', without any translation, with some kind of electrical attachment that plugs into the ear. The day when this happens is far away, but no one can say it will never arrive.

The second reason why the rules for computer communication will eventually be the same as those for voice telephone links concerns the exchanges that route signals to the correct destination. Computerised exchanges, once engineers learn how to make them properly (and because they are very complicated bits of machinery, this takes some doing), are faster and much more efficient than their analogue counterparts. In the latter, an electrical analogue signal sent along the wire instructs a moving contact to switch into a new position to set up a phsycial connection between an inlet and an outlet port of the exchange. The process, repeated several times within the exchange to make a connection for one call (which is why such hardware is said to use a 'step by step' process), is slow and unwieldy. It is a relic of the mechanical age that is fast disappearing.

Computerised exchanges, on the other hand, perform the same function entirely electronically without any moving parts. The signal entering the switching centre is coded so that the system automatically knows where it should be heading and routes it to the appropriate electrical outlet. The process is to the old mechanical switching method what the telephone system itself is to writing a letter and sending it by post. What the former accomplishes by electrons moving through wires the latter achieves through muscle power and physical effort.

The old step-by-step, or Strowger, exchanges have been partially replaced by more modern ones called crossbar systems in some countries. Crossbar contains some electronics (in a rudimentary computer logic that works out which signals should go where) but is a long way from the fully electronic exchange. Of the new all-electronic system the world has yet seen hardly anything, with the notable exception of the US where some twenty-nine per cent of telephone connections are made electronically, forty-five per cent with crossbar and twenty-six per cent step-by-step. By contrast, the figures for Japan, France, West Germany and Britain are respectively 1, 80, 19; 1, 72, 27; 1, 0, 99; and 5, 6, 89. (Figures are for 1976, quoted from *The Modernisation of Telecommunications*, Post Office Engineering Union).

The new exchanges are quick, quiet and need hardly any maintenance once they have been properly developed. Furthermore, there are other savings that stem directly from the use of computerised parts. In Britain, various industry sources predict an average 'new generation' exchange will take up a tenth of the space of a mechanical system, require one thirtieth of the workforce to make it and possibly half the men to run it, and cost one-third as much.

The benefits, then, are considerable in cost terms. The new exchanges should also offer a better service. There should be fewer wrong numbers and better billing for telephone accounts (a computer tacked on to the main system sees to this). Other features such as automatic alarm calls and stored telephone numbers (where the dialling of a single digit connects an often-used number without the need to dial all the digits) will also become commonplace by subscribers being able to exert some computer control themselves.

Such attractions are not lost on the world's telecommunications administrations. Most countries in Europe are pushing hard to change over to a binary network for voice traffic in conjunction with their efforts to build purpose-built binary data networks. (Eventually the two will be merged in one integrated system handling both voice and data.) Most of the technical problems are almost surmounted. America is already well on its way towards the change-over. Britain's binary voice network goes under the enigmatic name of 'System X'. The Post Office is spending £170 million getting this ready for installation into many parts of the existing telephone network in the final twenty years of this century. The first System X exchange was installed in London in 1980 and by 1988–9 fifteen per cent of the network should be running according to System X principles, that is with both message transmission and routing carried out binary fashion. There will also be a battle by the big telecommunications companies of the industrialised world to sell all-electronic systems to developing countries. Many such nations have virtually no telephone system for much of their territories, and to step into the year 2000 with a super, not-too-expensive, all-electronic system will often be a temptation impossible to resist. Sales of telephone equipment to the developing world will account for about £1 billion of the £15 billion the world spent in this area in 1980, with the figures predicted to rise over the next decade by twenty per cent annually.

Computerised communication (and as you can now appreciate, voice traffic falls into this category) will first make its impact on the business person. But the most far-reaching effect is likely to be in the home. Fuelled by constant price decreases brought about by shrinking microelectronic technology and, perhaps even more important, by the fact that as energy prices rise it will be increasingly expensive to journey somewhere for information and therefore more cost-effective to search for it electronically, the home electronic information system will gain in popularity. It will encompass both voice and data links, allowing for home banking, shopping, recreation through puzzles and TV games, education through computerised lessons, even home work. We will explore how the computer will affect the home in the next chapter. But let us leave this one with these words of warning from a 1978 United States National Research Council report on *Telecommunications for Metropolitan Areas*:

'There are promising opportunities to expand, enrich, and simplify life through the new telecommunications, and to replace tasks that are boring, routine, and time consuming. Life could become safer, more productive, and more fun. It could be healthier and better organised. Travel could be reduced and energy conserved . . . By providing ready access to information and assistance of all kinds, the home communications-information system could improve the quality of life for a majority of people; but it could just as easily become a force leading to the break-up of society. Will so versatile a device cause citizens to isolate themselves behind a wall of electronics, reducing yet further their opportunities to develop those human contacts that are essential to a balanced existence? . . . And what use will be made of the time saved? Will it be used to take advantage of the new educational opportunities offered, or will it lead to further frustration and loneliness?'

Chapter 6: The introverted electronic society – chips in the home

The extent to which chip-aided gadgets for domestic use become accepted depends very much on how inward-looking a society we become. Microprocessor devices for the consumer are almost entirely meant for people's houses – and if people react against staying at home, then the electronic devices will remain on the shelves in the shops where they are sold, and the equipment makers will go bankrupt. However, all the signs are that these events are highly unlikely. One of the epicentres of the chip revolution will be in the home, which is where the increasingly affluent populations of the Western nations have shown they like to spend their time and, in stocking up with all kinds of domestic gadgets, much of their wealth.

The rise in living standards in the Western world after the Second World War showed itself at first in an extrovert tendency to go out and do things that required spending money. Dance halls in the early 1950s were packed for instance. Cinema admissions in Britain, to take another example, rose by forty per cent between 1936 and 1950. However, the coming of television heralded a change in this pattern. The TV set in the living-room meant that people could stay at home to fulfil their entertainment needs. The TV started what we may call the age of introverted electronic amusement – an age that is still with us today, although it is populated by many more products that offer home-based electronic amusement, a lot of them powered by chips.

A few statistics help tell the story of how much the UK relies on armchair electronic fixes for its relaxation. In 1951 only eight per cent of the population had a TV; in 1978 only eight per cent were without one. People in Britain in 1977 spent roughly the same on TV rental as they did on all other forms of entertainment. The spending on rentals rose from £347 million in 1970 to three times as much in 1977 (constant 1980 prices). Consumer spending on all electrical goods (including TVs and other entertainment systems but also appliances like washing machines and hairdryers) virtually doubled in the same period from the 1970 level of £1400 million. Cinema admissions, by contrast, gradually declined. The average person went to the pictures once a fortnight in 1950 – in 1968 he went only about five times a year. This, then, is the situation in which the cheap computer-in-the-home is likely to make its presence felt. Commenting on what could happen in the future, the US National Research Council report referred to in the last chapter forecasts the house becoming a communications and information centre with separate chip-run equipment for such things as energy management, education, accounting and communications (see figure 5). The report's authors do not expect complete systems to enter people's homes on a big scale overnight. Natural inertia against change is too great for this to happen. But there will be, instead, relatively slow changes with computers being applied to fairly ordinary appliances that are already part of the scene around the house. Thus there will be 'intelligent' radios that can record information received over the airwaves on paper tapes, or print bulletins of important messages. Also, homes will incorporate intelligent TVs that permit sophisticated forms of viewdata and computer-run games and education programmes, as well as home terminals, looking something like intelligent television sets but containing enough computer power to do household calculations or memorise messages.

Chips in the home will not appear only in entertainment systems. They will also be important in labour-saving domestic equipment like refrigerators, washing machines and vacuum cleaners. Already virtually every type of domestic appliance is represented in the number of devices that come equipped with

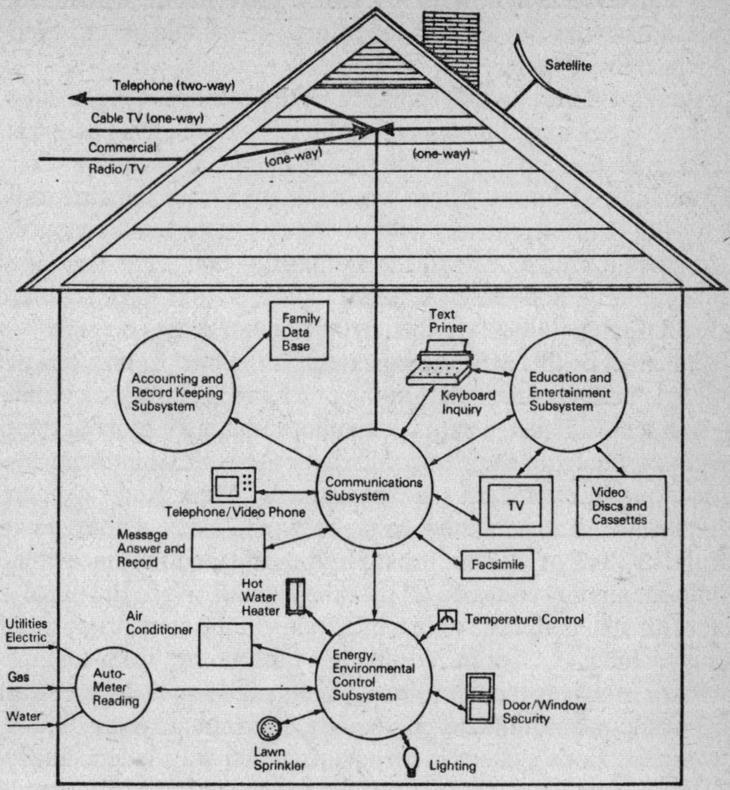

Fig. 5 Home Integrated Communications and Information System

Source: Telecommunications for Metropolitan Areas, US National Research Council

microchips. The chips' importance is that they enhance performance in some way. For instance they make available close control of a cooking operation in the case of an oven; they regulate a heater's output to match it to what is required to save energy; they can monitor the number of times a fridge's door opens to work out the quantity of heat it loses so that the internal frost/defrost cycle is worked out more accurately;

chips can even tell how many clothes are inside a washing machine to work out the correct volume of soap flakes required for a particular load.

The reason that people think it worthwhile spending money on these types of appliances to make their home lives easier is very much tied up with their desire to spend more of their leisure time at home. Faced with the uncertain, violent and increasingly unattractive world of the 1980s, people's reaction is to retreat into comfortable home-life, confident that the labour-saving and entertainment gadgets that they cluster around them will insulate them from the harshness outside.

The effect of this attitude repeated on the part of millions of people is difficult to assess. On the one hand, the outlook could breed a self-sufficiency with which people cater for most of their requirements from the electronic castles that are their homes – with the drawbridges firmly raised. Men and women might not find it necessary to go to work, even, if they have administrative or clerical jobs. High-quality communications would mean they could do all that they usually do in the office – react either to people or to bits of paper – in the comfort of their living-rooms. Children could be taught by long-distant teachers in much the same way. Even needs such as medical care could be administered, to some extent anyway, via a communications system incorporating two-way video, audio and computer links. This would encompass a kind of electronic doctor service.

These changes in society might be welcome – at least for some. However the changes could lead to an insular attitude on the part of people staring at the world through the medium of sophisticated electronic gadgetry and hardly touching it first-hand. Lifestyles based around the concept of the electronic home could also bode ill for the class of people who we can call the computer illiterates of the world. Such a class does not exist, yet – or if it does, nearly all of us are members of it. This is because there is no widespread familiarity with the machines. However the division into computer 'haves' and 'have nots' will begin to occur as more people become acquainted with how to control the devices that gradually muscle in on their lives. Life could become based around the computerised home terminal to such a degree that anyone who finds it impossible to learn how to use one – or has never been taught – would become an outcast.

For example, telecommunications links using computerised terminals could become the standard method of disseminating information from government authorities such as the post office, electricity boards and the health service. Certainly this smacks of a strange world reminiscent of science fiction but if it did come about, people who could not use the equipment would be at a big disadvantage. The NRC report picks up this point:

'Who can tell what the response might be when people are faced with a new telecommunications medium? Operating new telecommunications equipment might be many times more complicated than using a telephone or television receiver, and understanding the service might require substantial education. Unfortunately . . . a significant fraction of the metropolitan population suffers from illiteracy and learning defects. Others are victims of anomie, the state of emotional depression and feelings of meaningless that may increase as mass-produced machines take over many previously individualised decisions and actions.

'It is an open question whether residents of the metropolitan areas (or residents anywhere for that matter) want such complex devices, even though they provide virtually instantaneous access to public services. Is there sufficient motivation to master their operation? Will it be more convenient, and perhaps more socially rewarding, to continue to seek out face-to-face assistance at government offices, say, or hospitals and schools?'

Though posing many questions, the report does not do much to answer them and this is its failing. However, there is admittedly still much to be learnt about the acceptability of new computerised communications' information systems. Indicators that, one day, governments could use them as a routine way of channelling information to the populace include the plans in France for Teletel, the French version of viewdata. Due to start in 1980 as an experiment among a test group of householders on the outskirts of Paris, Teletel will rely on a large number of information providers who will provide the text to be transmitted – among them are two French government departments that will offer details about heating allowances and schemes for handicapped people. In the US the Department of Agriculture has its 'Green Thumb' viewdata service that provides information for farmers. And in the UK, several bodies linked with government organisations are among the information providers with Prestel. For example, the Health and Safety Executive gives details, through the service, of safety legislation.

Consumer-orientated electronic equipment accounted for world sales of £7 billion in 1976, or fifteen per cent of the whole market for electronic goods, according to the consultants Booz, Allen and Hamilton. The figure will probably rise to £12 billion by 1981 (at constant 1976 prices), a growth-rate of twelve per cent per annum, which keeps pace almost exactly with the overall growth in demand for all types of electronic goods. Typically, the semiconductor part of the equipment accounts for about five per cent of the price. Companies that make chips do not usually sell consumer gadgets. There are exceptions to this – for instance National Semiconductor and Texas Instruments make calculators and other types of goods aimed at the consumer, and IBM, the world's biggest computer firm that makes plenty of chips for its own use, is now getting interested in selling home computers and domestic video equipment. But normally the chip firms sell their products to the equipment producer (for instance a washing machine firm) which then incorporates them in the final piece of domestic apparatus.

According to the US magazine *Electronics*, the US alone

purchased some £6 billion worth of consumer electronic goods in 1979. The breakdown is set out in Table 2.

TABLE 2: Spending on electronic goods for consumers in US (1979)

consumer equipment	*sales ($ million)*
TV	4197
audio/radio	4142
microwave ovens	1500
intrusion alarms	255
smoke detectors	85
telephone answerers	35
electronic games	840
calculators	665
electronic watches	566
digital clocks	88
other chip products	1668
total	14041

The first consumer product to incorporate chips was the device that made most people aware the chip revolution had begun – the pocket calculator. In a very short time the calculator rose from sales of virtually nothing to command a leading position among the world's electronic products. According to Ernest Braun and Stuart MacDonald in their book *Revolution in Miniature*, the first electronic calculator was produced by a UK firm, Bell Punch, in 1963. Later in the 1960s, more sophisticated electronic circuits became available as integrated circuit technology matured. But the world had to wait for the invention of the microprocessor in 1971, the first chip that could be programmed, for the basic building block for calculators that would make the devices both flexible and inexpensive. Meanwhile, electronics firms from Japan, and to a lesser extent the US, took the initiative. Calculator companies produced thirty-four million of the gadgets in 1974 and ninety-one million in 1978. The price of a standard calculator fell from $100 in 1971 to $11 in 1978. By 1975, the US was a calculator nation: one person in six owned one.

Digital watches followed a similar pattern. They are powered by a tiny chip that keeps track of the oscillations of a quartz crystal, converts them via its logic into hours, minutes and seconds – and displays the result, via a mechanism such as liquid crystals, for the benefit of the watch's owner. In 1970 few people would have predicted that the traditional Swiss watch, powered by its intricate clockwork mechanism, would be all but swept aside by a new, rival kind of instrument. The Swiss watch industry certainly never foresaw this possibility – to its cost. It had to endure several years of slump as the digital watch registered very heavy sales in the mid 1970s. The sales figures jumped from 250,000 worldwide in 1973 to three million in 1975 to an estimated fifty million in 1980. In that time prices plummeted from $250 to $40 or less.

Now that sales of watches and calculators – the two bastions of the chip novelty world – are beginning to fall off a little, the chip industry is on the watch for the next boom gadget. For the time being, TV games and toys fill this slot. The former are chip modules that plug into a TV set, their programme residing on cassette tape, to control the picture elements on the screen in such a way that they simulate, for instance, a Moon landing or a tennis match, allowing the user to change the picture elements with control buttons. In this way he can play a series of games that range from the simple to the sophisticated. According to the Economist Intelligence Unit, world sales of such video games grew from 3.5 million in 1976 to 11.5 million two years later. The US soaked up ninety-seven per cent of the devices in 1976 and just seventy per cent in 1978 by which time the rest of the world had woken up to the TV game craze.

Chip toys range from model tanks and other vehicles, that a child can program to carry out complicated manoeuvres, to hand-held gadgets, looking like bigger versions of calculators, with which people can play electronic versions of such things as chess, Scrabble or Mastermind. Such 'games' often have an educational flavour to them – manufacturers find this helps sell the things. An example is the $55 'Speak and Spell' game from Texas Instruments, a company traditionally with an eye to the consumer market which sells a lot of calculators. The device asks its user to spell words that it speaks out for them with the help of a voice-synthesiser chip. The value of hand-held chip games from all manufacturers increased from £10 million in 1977 to £180 million in 1979.

What kind of gadgets will come next in the consumer goods world? 'I don't know what the future will be,' confesses Jim Huffhines, microprocessor marketing manager for Texas Instruments. 'In 1975 it was not obvious that toys and games would become a big market for chips. New things will come on to the scene. That has happened in the chip business since Day One.'

Much of the new equipment in the home will cluster around the TV set and the telephone – devices that may have seemed strange a few years ago which most people now accept unhesitatingly. No exception to this pattern is the video recorder, a device which Japanese firms such as Sony and JVC, almost single-handed, have made a roaring commercial success in recent years. The recorder plugs into a TV set and with the aid of its controls – more than likely to incorporate a microprocessor – turns itself on or off at the correct times to record a favourite programme while the TV's owner is otherwise occupied. Alternatively, a video recorder can be a 'stand alone' version which, rather like a gramophone, 'plays' cassettes or discs of video with the pictures flashed up on a screen. In 1979 about 90,000 people in Britain owned a video recorder: the number was expected to at least double in the following twelve months. Sales of video recorders in Europe were expected to double between 1979 and 1980 from the figure in 1979 of 225,000, according to a survey by Mackintosh Consultants.

Home computers are an obvious consumer gadget. However they look like being popular only among those seriously committed to computers. You need a lot more skill to handle one of these than, say, a chip-controlled washing machine.

Nonetheless, recent history has shown just how many enthusiasts there are. Home computers are likely to be popular among hobbyists who have something of the gadget-crazed mentality of the radio ham or, to a lesser extent, the model railway owner or the custom car nut. The further attraction of the home computer is that it can actually do something useful. As well as providing a way of passing the time with the help of computer games which are plugged, in cassette form, into the machine, the device can work out calculations, store personal messages and act as a filing system – even write letters using a word processing programme.

Statistics for home computers can be a bit confusing because many people who buy the devices – they cost anything from a few hundred pounds upwards – are not strictly home enthusiasts. Rather they purchase the machines to help them at work or at school. The world market for home computers in 1978 was about £230 million. The people using them can be split roughly 7:2:1 among business users, true hobbyists and people at educational institutions.

A question people quite understandably ask about microchip-aided products is 'What use are they?' We can help answer this one by considering microwave ovens. These are a relatively new form of oven suitable for today's living habits where a rapid 'warm-up' of a meal with microwaves is often more acceptable than proper cooking – even though the food is likely to be less pleasant as a result. Microwave ovens were virtually non-existent in the 1960s before the microprocessor age began: now they sell at something like four million a year, mostly in the land where fast food reigns supreme, the United States. The ovens were specially suitable for fitting with new forms of controls because there was little resistance from manufacturers. The ovens required totally new production lines anyway so the idea of microchip controls was not as novel as it might have been to other equipment makers. As most gadget-crazed American households would be able to tell you, many kinds of microwave ovens permit control over cooking that is unheard-of in conventional ovens. For instance, you can tell the oven exactly how much heat you want it to impart to the food, tell it to defrost some frozen peas or beans, or order it to cook only to a certain temperature and then turn itself off. Whether the device improves the taste of the food is open to question: but one thing that cannot be disputed is that it's great fun to play with.

What are called intelligent thermostats are another type of device that will benefit from microchips and appear in people's homes in large quantities. These are pieces of equipment that regulate a building's heating controls according to a great many variables such as outside temperature, barometric pressure, the preferred comfort level of people inside, and occupancy trends (for instance how often people go out or open and close windows). About half of Britain's energy is consumed in buildings. The country's Building Research Establishment estimates that thirty per cent of this energy could be saved by fairly simply methods, such as the better control of appliances that would follow from microchip monitoring. This would save something like £2.4 billion annually. (The figures include energy consumption in industrial and commercial buildings which account for a considerable part of total building energy usage.) The intelligent thermostats could soon start to appear in shops as add-on devices to existing appliances and costing as little as £50.

While considering consumer-spending on microchip devices, we must not forget the car driver. After all, a car for many people represents a moveable extension of their homes – so car gadgets fall within the category of items covered in this chapter. Chips for motor vehicles will soon become very important. The US car industry produces about ten million vehicles annually, about a quarter of world output; and by the early 1980s virtually every one will contain at least one microprocessor supported by dozens of ancillary chips. The reason for this is that tough government rules on exhaust emissions and fuel economy have forced the American vehicle firms to tighten up considerably on the traditionally inexact method of controlling engines. This has led to the use of a tiny computer to monitor a dozen or more factors affecting engine performance – for instance the temperature inside the combustion chamber or the position of the crankshaft. The microprocessor then makes sense of the readings and instructs other parts of the engine to react accordingly, to make sure that fuel economy and the quantity of poisonous gases in the exhaust output are kept at optimum levels. The electronic systems that will be fitted to virtually all America's 1981 cars will cost \$300 to \$400, including both the electronic controls themselves and the relevant sensors and actuators that feed them with information or react to signals which the controls produce. Of this price the microprocessor chip accounts for just a few dollars, so proving the point once more that microprocessors aren't a lot of use unless they are surrounded with a host of other devices for both supplying them with data – acting as the chip's 'eyes' and 'ears' in other words – and for carrying out what they decide.

On top of the applications in regulating engines, microprocessors have already started to appear in the dashboard instrumentation of motor cars. They control panels that supply snippets of information that either amuse or assist the motorist. (Microchip aids for the car are in many ways what calculators or personal computers are for the home: their use depends very much on what the owner wants to get out of them.) The 1980 Lincoln Continental luxury cars made by Ford contain what the company calls an electronic message centre. For most of the time this functions as a digital clock, but it can also be instructed to tell the driver how far he is from his destination, how long it will take him to arrive. With a bit of ingenuity it could probably even tell him what the weather will be like when he gets there.

As a further bonus, the unit flashes up messages on to a small screen that informs the driver if anything in the car goes wrong. In this way the driver will be alerted to engine problems as well as little things like a door that has not shut probably or a light bulb that has failed. A similar 'problem alert' device will be included in General Motors' 1981 cars. This is not quite as sophisticated: the display flashes up a numerical code which the motorist will not be able to decipher immediately. Rather, it will tell him to drive to the nearest garage where a mechanic will 'unscramble' the code with a book of instructions. The procedure will be effective if the garage hand knows how the device works – instructing car mechanics in the arts of computer engineering will be a difficult job and something that the big car firms like General Motors are more than a little worried about.

The most sophisticated dashboard devices will for the time being appear only in the most expensive cars – the Lincoln Continental for instance costs about $14,000 – but if consumers take to them strongly enough they will soon be incorporated in more mundane vehicles. Although they have been slow compared with the US companies, European car firms will begin incorporating chips in their products as the 1980s progress, in step with new anti-pollution laws that compel strict engine controls. Japan, which already has tough anti-pollution legislation on a par with what exists in the US, already has a higher proportion of cars than any other country that contain electronic controls of some kind to regulate carburettors.

Lest you think that any of this chapter has been the least bit fanciful – many people are inclined to scoff at the idea of the computerised home – it is as well to point out that one such house already exists. It is an unusual building just outside Phoenix, Arizona, which is controlled by five microprocessor-based modules. The most fanciful part of this structure is its name: it is called the Ahwatukee house after an Indian word that means 'shining house of dreams'. The rest of it is extremely practical. The computer system monitors who comes in through what is called a keyless entry system: instead of keyholes there are little buttons on each door which people press if they want to enter. If the computer recognises an individual's code, it lets him in. If it doesn't, it's hard luck, and the person probably gets a rude message into the bargain, by courtesy of a voice synthesiser unit which the house's owner can program to give a series of computer-driven spoken statements.

The microprocessor system monitors several other functions such as energy consumption, house security (it coordinates fire and burglar alarms) and eventually it could even arrange for the plants to be watered at the right times. The system also turns on and off the lights and provides electronic message pads for the house's occupants. People 'write' their messages using a keyboard, and a screen displays them when they are required. In this way, the computer could act as a store point for diary entries or details of the family's favourite recipes.

Built by Presley of Arizona, a property development firm, and fitted out with the controls by electronics group Motorola at a cost of $30,000, the house will be open to visitors until 1982 when it will be sold. Motorola thinks the house could be a good testing ground for the type of home controls that could become common over the next few years. Perhaps the most significant thing about the building is that almost anyone would be able to handle the computer equipment. The householder would not need to be a computer programmer. 'He would need a couple of days to set the system up,' says Pat O'Malley, one of the engineers who designed the apparatus. 'After this, unless he changes his mind about what he wants the system to do, he wouldn't have to bother himself with it at all.' That is how the computer will affect the householder – it will be cheap, commonplace, and largely invisible.

Chapter 7: Computers at work

'The development and application of the automatic electronic computer – commonly known as the 'electronic brain' – for the integration and control of office procedures and manufacturing processes is something new, beckoning us towards the electronic office and the automatic factory of which we can already see the outlines.' So starts a little booklet on automation that the British government had the prescience to publish – in 1956. Automation, under computer control, of the processes involved in industry and commerce has clearly been around for a considerable time. So why make a fuss about it now?

The reasons are twofold – they concern costs and experience, the latter breeding a third factor, sophistication. Together these explain why things aren't what they used to be in the automation business.

First, although the ideas for harnessing computers to run offices and factories are not new, they were never implemented on much of a scale when computers were expensive. Consider the example of the Lyons Electronic Office, called LEO for short, which was the forerunner of ideas we have today for bringing computers into clerical work. LEO was the brainchild of the catering company Lyons, which had no interest at all in computers other than to think – with commendable foresight – that they could help in business. So Lyons engineers built LEO based on the ideas of the Cambridge EDSAC (see Chapter 2) and it started work in 1953. The machine was hailed as a marvel. It was calculated that LEO could do the job of six hundred full-time clerks if put to work on payroll calculations eighty hours per week all the year round.

However in 1953 LEO cost about £100,000, which at 1979 prices was worth approximately five-and-a-half times as much. In other words the computer was hideously expensive. Great though its hypothetical savings were, in cost terms it could not be justified unless put to work on very big jobs where these savings became real. Similar arithmetic applied to other computers of LEO's generation. Today the cost-to-savings ratio is very different because of the reduction in prices of computers. A minicomputer of LEO's power can be purchased for £15,000 or so; in other words almost one fortieth of the price. (What is more it fits into a box the size of a filing cabinet, whereas LEO took up a room.) Thus computers are economically feasible today for many of the applications for which they were not in the 1950s.

A further economic reason why computers are more attractive for displacing labour is that labour costs have risen at a greater rate than prices in Britain over the period 1955 to 1980. In that period average wages went up seven-and-a-half times while prices rose five-and-a-half times. Hence assuming that all other factors have remained the same, the proportion of labour costs in the average company's expenditure will be higher in 1980 than it was in the mid-1950s: and so there will be more incentive for shedding staff by introducing labour-saving computers.

The second factor that has made computers more of a force today can be simply put down to expertise in adapting them to the areas of business activity for which they are suitable. For one thing, software advances mean that computers can be fairly easily changed from one job to another. Today it would not be too much trouble to convert the old LEO from its role calculating people's wages to, say, co-ordinating billing information in a hotel or sorting papers and memos in a filing department. These kinds of advances make computers flexible, and therefore attractive to companies which want to harness their machines to several different types of operation.

The example of computer-controlled machine tools is relevant when considering improvements in computers' flexibility. The first such tools were controlled by lengths of paper tape in which holes were punched to represent numbers, the numbers instructing the machine in which sequence of operations to perform. In this way a lathe, or a milling machine, can be made to operate without a person having to handle it, except to change the tape now and again. However a complex operation could require a tape perhaps 100 metres long. Imagine what would happen if someone wants to alter the program in a small way, perhaps in a section right in the middle of it. The person would need to 'play' the tape back in the manner of an audio recorder, and mess about changing the pattern of holes: he might decide the time and effort is not worth the bother. Today, however, solid-state memories have advanced to the point that one chip can hold the program: and within this it can be altered very easily, by playing with buttons on a keyboard, so that one tool can carry out a lot of different operations.

To sum up, the fact that computer devices have become smaller and cheaper – and that engineers are more adept at using them – makes it fallacious to talk about the impact of today's computers in the same breath as those of the 1950s. We are talking about the same animal, to be sure, but a totally different species. This is why computers' effects on industry and working habits are likely to be very different now from when they were first introduced, when, it is widely agreed, many of the forecasts made about their unsettling impact proved strikingly inaccurate.

Today it is virtually impossible to think of any area of business activity which is not affected, or has no possibility of being affected, by microelectronic chips. According to the consultants Booz, Allen and Hamilton, world sales of electronic equipment for industrial and commercial use accounted for twelve per cent of all electronic goods in 1976, representing sales of £6 billion. The figure is due to climb to £11 billion by 1981. An increasing proportion of this electronic equipment is coming under the control of intelligent control systems, invariably incorporating microprocessors.

We shall consider the subject of computer-controlled

business machines in the three principal areas they affect – the office; manufacturing and processing industry; and in service industries such as shopping and banking.

The office: end to paper problems

The automated office of the future will be centred around computerised devices such as the word processor, a microprocessor-based typing system that we have already met (Chapter 5). Such pieces of equipment mark an important discontinuity in the way mankind organises itself. From the time the Egyptians found they could write on parchment, made from river weed, paper has been increasingly important as the device for transmitting one person's thoughts to another. Electronic devices for the office herald a radical change in this pattern: their impact will be to cut down on paper.

The world of the office worker is basically one of handling little pieces of paper. In the US alone seventy billion documents and memos flow to and fro among people in offices. There is a Laurel and Hardy film in which Oliver Hardy is shown in his occupation as a clerical worker. He has an unusual filing system – a huge pile of untidy documents on his desk which no one else in the office can sort out. Asked to find a particular piece of paper, Ollie with amazing accuracy lifts up the pile at the exact point he wants, withdrawing the document and shoving the rest in place to make the same disgusting mess as before. Other more tidy-minded people in the office grimace at the nastiness of it all. This is a situation you would never find in an electronic office.

Figures and words are committed to paper in the electronic office only when really necessary. The rest of the time they reside neatly stacked away inside electronic memories in computer-like apparatus, to appear at the press of a few buttons in disembodied form on display screens. There they can be moved around, edited, filed away or sent in letter form (using binary communications links) to other offices either in the next corridor or hundreds of miles away. Only if someone decides he must have a copy of a document will it be printed for him – in a few decades' time printed pieces of paper could be so rare that they are collectors' items.

The introduction of the word processor is the first step towards this idea of the electronic office, which it must be said is little more than a vision at the moment. But the rest of the jigsaw pieces that make up the complete picture are gradually falling into place. Many big companies in the US and the UK already employ word processors to take the place of ordinary typewriters. The advantages, outlined in Chapter 5, are that the typist takes less time to correct mistakes – he or she just presses a button instead of having to worry about things like whiting-out fluid – and can use the machine's computer power to edit or store messages.

The benefits lead to labour savings. Typically, one clerical worker with a word processor can do the job of two or three people with an ordinary typewriter. There have been few cases of redundancies brought about by word processing equipment: what usually happens is that a company which introduces the machines finds there is less need to hire new staff if it is expanding; or, if it is not, it reduces the number of employees slowly, by natural wastage. The British clerical workers' union APEX has estimated that 20,000 job opportunities have been lost by this kind of process over the last few years of the 1970s.

In its booklet *Office Technology – the Trade Union Approach*, the union points out the dangers of increased unemployment due to word processors. Yet it also comments on the advantages to the workforce that come through using the machines. Some office jobs can be made less tedious with the time spent on mundane typing jobs made available for more interesting occupations. On the other hand, there is the possibility of typists being relegated to the position of machine-minders constantly fed – possibly by audio links – with reports and letters to write on the machine. Diverting activities such as filing or taking shorthand notes could disappear. APEX advises: 'The detrimental effects of office automation . . . can be overcome if the office is treated as a total complex organisation and the technology designed and applied to suit the workers who use it. This will necessitate the redesign of jobs, and experimentation with new systems of work organisation such as small work groups.'

Further ahead on the road to the fully automated office is the concept of what is called the integrated work station. At the moment, such systems exist as little more than ideas although several companies are struggling to turn them into reality. In Britain, among these firms are Nexos, which the government set up as a subsidiary of the National Enterprise Board in 1979, and the better-established electronics trio of Plessey, GEC and ICL. In the US, IBM and Xerox have done a lot of research work and other big firms in the £23 billion per year world-wide computer business – such as DEC, Honeywell and Hewlett Packard – are keeping an eye on developments.

The integrated work station would look similar to any kind of intelligent computer terminal – that is, it contains a screen and a keyboard and has a comfy seat in front of it. At the touch of a few buttons, the user could display on it text that he would normally have to shuffle through drawers or filing cabinets to find; the unit would contain a computer store for associates to leave messages, in either textual or audio form; and the executive could summon his secretary, or anyone else, through the telecommunications system built into the machine. (Messages would travel as binary digits so that computerised devices in the office can talk to each other directly.). The work station would double up at any time as a viewdata terminal, with which the user could summon information stored in publicly-accessible data banks anywhere in the world – using the kind of communications links we talked about in Chapter 5. To write letters or memos, he would probably type them himself with his keyboard and screen (there would be nothing demeaning about 'do-it-yourself' activities like this in the office of the future), leaving only the long and more complicated documents to his secretary. The work station would possibly be linked to the firm's ordinary data-processing computers that deal with payroll, personal records and so on, so that people at a keyboard (if they have the correct passwords) can at any time get access to the details of the company's operation.

Costing no more than word processors today – that is £10,000 or less – such devices could be common in the offices of all but the least successful companies in the 1990s. Companies will probably allocate them to staff in much the same way as today they assign people to desks. Not just the top executives would get a machine; the more lowly grades of staff would have one each as well. This is so that people higher in the hierarchy can tell them what to do using these efficient communications links to increase productivity.

Office computers in the 1990s will probably be able to talk and listen. Already some of today's computers are capable of fairly rudimentary speech. The machines are programmed so that their logic circuits trigger audio signalling equipment that broadcasts sound waves representing the sound of the human voice. Such pieces of equipment – which can be put in chip form – are called voice synthesisers. However most of the computer talk today is a long way from being perfect: the 'speech' is like that of a Dalek who badly needs elocution lessons. Further research is needed.

Progress with computers that have 'ears' is even further away from producing an acceptable product. To make computers that can identify words, and turn them into signals that their logic circuits process, engineers need to devise very accurate sound-metering equipment that splits up the sounds of a voice into the basic 'phonemes', or building blocks, of the spoken language and then compares these with a 'computer vocabulary' in its memory to work out what the words mean. There are some computers that can recognise perhaps 100 or so words – and these only if spoken with a certain pronounciation – but there is a long way to go before they become acceptable for use in environments such as the office. The great attraction of seeing/hearing office machines is, of course, that there would no longer be a need for input and output devices such as keyboards and screens which people have to learn to handle; to communicate with computers would be no more difficult than doing the same with a person: computers would then have overcome the final barrier to complete acceptability.

The automated factory

In a normal factory – indeed in any process in which something is made – there are several stages in the manufacturing process. First the object or product has to be designed (or defined in some other way, if for example the product is a chemical mixture of some sort) so that the factory's workforce knows it is making something that will meet requirements. Second comes the job of putting the object together, which typically involves activities such as assembly, metal-working, painting, joining or chemical processing. Third the object has to be inspected to make sure it is the right shape, weight and so on, and otherwise meets the specifications laid down in the design. These three stages are fundamental. In the world of the late 1970s it was theoretically possible to leave each one of them to the control of a computer.

However, you are unlikely to visit a factory and see computers overseeing all three stages in concert. Indeed, it is doubtful whether such a place yet exists. One reason for this is that it would drastically go against past practice. Most factories will introduce computerised automation only slowly here and there. Neither workforce nor employers want it particularly badly, if for no other reason than that too much technological progress goes against the human psyche. The second reason is more important, and more simple: it is very difficult to make integrated automated factories work. Giving a computer one part of the manufacturing operation to worry about is easy enough. But connecting a lot of computers so they each deal with different stages – and then link up effectively – is a different matter. There are a lot of technical problems. That does not mean, however, that no one is trying to solve them.

Researchers at West Germany's Institute of Machine Tool and Production Technology in Berlin have built prototypes of a fully automatic factory in which the main elements are a series of computer-controlled machine tools that work on individual parts; robots that put the parts together and test for quality control at the end of the process; and overseeing computer systems that continually receive information during the manufacturing processes and make adjustments to, for instance, the number and type of components contained in the factory's stores that are fed into the production lines. The link between the manufacturing that takes place on the factory floor and the design and storage areas of the factory helps workers make alterations to devices they are working on: for instance a factory hand could instruct the computer looking after the stores that he requires a different kind of component for an unusual model. This would make sure that the correct one is selected and sent on its way to the production line. In this West German vision of the future production unit, people are still needed to man the individual tools. But there are not so many of them as would be required today, on account of the computer system that handles many of the links between different parts of the manufacturing process. Foremen and other 'overseer' types of worker will probably not be required. Added to this, the extra automation of the machine tools due to the computer system will probably mean one person can handle two machines instead of two, decreasing further the amount of labour needed.

Other engineers in Sweden and Japan are working on similar model factories in which less people are needed to do the work. The big technical problem that stands in the way of complete automation is computer software. It is all very well designing a robot arm that will manipulate things under computer control; or a machine tool like an automatic lathe that shapes things; or a microprocessor-controlled instrument rack that keeps track of a dozen different variables, relayed to it by sensors, and feeds the information to actuators that react accordingly. But even if the software to implement these jobs can be developed (and this takes a lot of time and trouble) then it is a fearsome task to devise ways of relaying instructions from one device to another so that they are continually in touch in a fully integrated operation.

To try to make this happen, engineers must develop suitable 'protocols' (software 'passwords' transmitted along communications links) that ensure each computerised device understands what the next one in the chain says. Otherwise a ring of devices all transmitting commands to each other – in which, for example, a controller unit says to a robot the equivalent (in binary signals) of 'grab the next pellet a little more to your left or else you'll miss it and that'll make us look right idiots' – could turn out to be similar to a shouting match in which people scream messages to each other in different languages, an electronic version of Bedlam.

Communications, then, are the problem. Many of the technical ideas being applied to find solutions are similar to ones put into practice for improving ordinary computer-to-computer links of the sort we met in Chapter 5. However there has been considerable progress in the individual parts of the factory processes. Let us now look at each stage in turn, starting with design. Computer-aided design is now a reality in many industries. We introduced it earlier (Chapter 3) in the discussion of how computers help in the design of integrated circuits. Basically CAD cuts out a lot of time that is normally spent drawing models of things that a factory wants to make.

Let us consider the example of an engineer who is designing pipework for a power station. He sits at his computer terminal (the computer to which he is linked must obviously have a suitable CAD program in it), and feeds into the machine, through the keyboard, data such as the characteristics of the fluid that the pipes will have to carry, how far it will have to travel, the bends the pipes will have to negotiate and a lot more things besides. Providing the software has been designed cleverly enough (you can, of course, get a computer to help you do this as well) the machine will plot out on the terminal a drawing of the required piece of pipework, possibly giving the optimum dimensions separately on hard copy.

There are many other examples of industries in which CAD is common. One of the biggest businesses to use CAD is aircraft manufacturing. At the Boeing aircraft company in Seattle, Washington, for instance, designers put their ideas into computer code and store them in computers for use in designs in succeeding years. The ideas form a library that engineers in years to come can call upon. As the company explains it, it's a way of retaining the skills of designers who have retired. In one of the Boeing projects, on the 767 aeroplane, 1,300 engineers all design different parts of the aircraft. The computer links between the workers lets them keep in touch with the way an alteration to one part of the plane will affect others – for example, how a change in the design of the fuselage will affect the shape of the wings.

The second stage in our factory process concerns making the objects, once they have been designed properly. This is where computerised machine tools enter the scene. The specification for the object in question – for example a cog for a gearbox for a car – can be translated into computer code and stored in a correctly programmed computer. From here the details are channelled to the tool, the automatic equivalent of a manually-operated machine tool such as a lathe or a milling machine. Instead of a factory operator moving the controls to make the machine move in a set sequence to make the part, the computer does it. Feedback signals from sensors on the machine tell the computer what progress is being made on the job so that, if necessary, the processors change the operating procedure to allow for problems, such as a piece of the part becoming stuck in the tool.

The first machines of this kind were fairly unsophisticated. They used paper tape as the storage medium for the computer code. Appearing in the 1950s, these so-called numerically-controlled machines were unreliable, slow and expensive. So they took a long time to catch on. In 1966, the annual world production of these machines was still only 2,500. This figure rose to almost 20,000 in 1978. Both software and hardware advances have led to the tools becoming more useful today: the incorporation into the machines of solid-state memory is the most significant step forward.

Once individual components have been machined, for example by cutting and shaping by a lathe, they normally have to be joined or in some other way fitted to some other component. In the case of a cog in a gearbox, the cog would have to be fitted to the rest of the box. In the world of the computerised factory, this activity is the province of robots. Robots are emotive things because a lot of people, having read about them in their youth in *Dan Dare* comics and the like, expect the machines to look human. What is more, people often expect robots to muscle in and take them over in some way. Robots are, after all, always supposed to be 'taking over' something, who or what normally being left to the imagination. In fact robots should be thought of as adjuncts, or peripherals, to a computer, and which carry out human-like actions. So they don't necessarily look like people at all. Admittedly, there are some humanoid robots, but the most common in industry are long, jointed pieces of metal that have several degrees of freedom. In other words they have the equivalent of wrists, elbows and other such joints. One end of the 'arm' is pivoted and linked to a computer: the other is the 'hand' (it can contain the equivalent of fingers) and carries out whatever needs doing.

There are about 8000 programmable robots in the world. This definition includes robots that can do more than one task – brought about by changing the software in the computer – and excludes the many thousands more similar machines that are 'single purpose' devices. These are usually controlled by paper tape that is difficult and time consuming to change. Half the world's programmable robots are in Japan, 2000 in the US and the rest in Europe of which only about 100 are in the UK. There has been much moaning and groaning about this last figure, with the pundits forecasting that Britain's industrial future will be bleak unless the nation does something about raising it – but we will return to this argument in Chapter 10.

The most advanced robots can be programmed exactly as one would an ordinary computerised machine such as a word processor or a desk-top data processing machine. In other words, the user types statements in computer language on a keyboard, the robot doing the job specified by the terms in the program. Robots cost anything from £15,000 upwards: the electronic part of them is fairly small, so they are not getting all that cheaper. However, they are getting more sophisticated, measured in terms of the number and complexity of operations that they can carry out.

Most robots in the late 1970s were used for what are regarded – in robot terms – as fairly easy jobs, for instance paint spraying or welding. Here the tasks are repetitive and simple to learn. Once the robot has been taught, it can sit in place on the production line and do the same task over and over again as components come past at regular intervals. For instance, to paint a particular part of a car body, exactly similar car bodies are passed in front of the robot, say, at the rate of one a minute. The user has to tell the machine to paint a certain area in a particular manner (these things are specified in terms of positions fed into the computer's brain) and to repeat the process every sixty seconds.

Robots today are now capable of more complicated activities as well. For example they can do things such as screwing in light bulbs on a car's instrument panel during assembly. Unimation in the US, in conjunction with General Motors, has developed a robot called PUMA that does this job with an accuracy of 0.1 mm. General Motors is developing PUMA to work alongside human operators at one of its car factories. Two other US firms. Cincinnati-Milacron and

Westinghouse, are developing similar robots.

The biggest problem with the machines is how to integrate them with the other activities in the factory. At the moment robots have to be 'spoonfed' like a baby. They work extremely well, in harsh environments and without flagging, but only if they always do exactly the same action. Once programmed, there is no way of altering this. So if a part is delivered to a robot on a production line a centimetre or so away from the position at which the machine expects it, the robot will act as though the part was in its normal position. The result could be, for example, a car body with paint going into the vehicle's interior by accident; or welds put on the wrong part so that a bumper is a centimetre higher than it ought to be. It would be much better if robots could see what they are doing and take account of objects that come to them slightly out of position.

Researchers all over the world are working on this problem. A seeing robot must be fed with information – in the same way that people when they look around are presented with information via light rays. The robot, by contrast, gets its information either from TV cameras or position sensors such as pressure gauges which record where an object is by finding out whether it is exerting pressure on a particular spot. For seeing robots to work, information must be passed from sensor to computer around the 'feedback loop' with great accuracy. The technical difficulties are so great that seeing robots that can work on production lines do not exist – yet. They almost certainly will. If the cost of running these robots is less than paying people wages, then employers could start to phase out human workers, except for those activities where a human brain is indispensable.

Testing is the third and final part of the factory process in which computers can play a part. We can bracket in this category the controlling of individual processes in process industries such as chemicals manufacturing. Such control activities boil down to a constant checking of factors – for instance reaction rates and pressures and temperatures – which can be looked upon in the same way as testing of a finished product to inspect for flaws. Computers are useful in the electronical equivalent of dipping a toe into water to test its temperature: however, they can control thousands of 'toes' in lots of different containers of water, and, what is more, correlate all the readings in a sensible way.

The checker-computer in this role will be found monitoring a lot of different functions, either altering them to keep them in line with a planned mode of operation, or checking them against an optimum list of values to help in quality control. Because of the low cost of computing, factories can spread their microprocessor-aided testing modules to many parts of the production process so that each section has its own logic unit measuring what is happening. The units then probably feed the results to a big central computer which monitors the monitors.

Process controllers are useful in any industry in which there are chemical reactions, for instance in food processing, chemicals or the photographic business. The controllers, costing £10,000 to £20,000, sit in boxes the size of small filing cabinets and take stock of quantities such as temperature, position of valves, density, flow-rate to keep the reaction flowing in a specified manner to produce the correct products.

In Britain, one of the first industries to take advantage of this form of control was the food industry. For instance, you have a microprocessor to thank for a lot of the sugar that goes into cups of tea. Since 1975 a microprocessor system has monitored the processing of beet sugar in a British Sugar Corporation factory in Bury St Edmunds, Suffolk. And before long, loaves of bread will be made with the help of microprocessors: Baker Perkins, a maker of bread-making machinery, plans to incorporate the little devices in new dough-mixing apparatus that should appear in 1981. Another very different industry where microchips will be employed is mining. The National Coal Board has developed a chip-aided coal-cutting machine that monitors the progress of a cutting blade as it travels along a coal face, transmitting instructions to a driving motor so that the blade cleaves out the coal exactly where required. The machine could become standard in Britain's pits in the mid 1980s. Once again it is an example of a computer brain taking into account a lot of different variables and doing a much better job than a human could.

The car industry provides a good example of computer applications in both monitoring and quality control. Chrysler in the United States has twelve types of microprocessor-aided systems in its factories. These do anything between checking thicknesses of plastic on car components to controlling a test cycle for instrument panels to evaluate whether the panels' lights are working properly. General Motors, the world's biggest firm, with a turnover roughly a quarter of Britain's gross domestic product, is not left behind. In one of its US plants, a $1.5 million monitoring system controls some 2000 production units, each of them manned by one operator, so that at any time the factory managers have an exact record of what is going on at shop-floor level. The system also keeps track of the factory's energy consumption. It monitors 450 sensors fitted to appliances such as heating and ventilation equipment. As well as saving an estimated $1 million in fuel costs in eighteen months, the system should cut ten minutes of 'non-productive' worker activity per man-shift. That could mean, in a workforce of 2000, that the same amount of production is carried out by thirty fewer people.

Better control in processing and checking does not, however, necessarily result in lay-offs. More likely, the result is a better product because the factors that influence its manufacture are better controlled and better understood. After all, in very few factories are the microprocessor controls likely to replace men and women who sit with pencils and notebooks jotting down the details of a chemical reaction. The chances are this kind of monitoring will just not be done, to the possible detriment of the product.

Computers at your service

Service industries – where people get paid not for what they make but for what they do – constitute an increasingly important part of life in the Western world. People who run shops and banks, bus conductors, insurance men, hotel workers, ticket clerks, firemen and policemen, shoe shiners, railway drivers, nurses and doctors, maintenance operatives, dustmen – these are all service workers, and there are a lot more besides. How will cheap computers affect them?

In some ways, service jobs will be among the least affected by computers. Not even the cleverest computer can empty your dustbins, provide medical treatment, clean your house, cut your hair or tidy up the public lavatories. (If anyone says 'leave it all to robots', you can afford to scoff at them. robots that can do these things won't be around until the end of the century at least.) But those service industries that rely on meeting people's needs by transferring information will be affected. For information, as you should know by now, is something that computers love to get stuck into. They are seldom happier than when given a truck load of information to sort out.

A lot of shopping is about passing information between the customer and the store assistant to communicate what the former wants to buy. Given good enough computer links (and good graphics capabilities so people can see what they're buying), a lot of this interchange of information could take place between the customer at home and the shop worker perhaps hundreds of miles away. Banking, too, is much concerned with relaying information from one person to another. After all, money only evolved because it is a way of informing one person how well-off another is. Money became popular because it is a lot easier to inform people about your degree of affluence by handing over cash than by giving them a lot of goods such as a herd or two of sheep or goats. In the evolution of other ways of replacing goods-transfer as a method of buying things, cheques became useful on account of their convenience. The next step in this chain of advances could be that everyone owns a portable banking terminal. With this, monetary transactions are carried out, not by people writing out messages on bits of paper (cheques), but by doing the electronic equivalent. They inform their bank electronically that a certain amount of money should be transferred from their account to someone else's. The advent of credit cards (in the US you hardly need money at all but buy virtually everything from petrol to a hotel room with a card) is a step in this direction. Before long, if these trends of shopping and banking go on to their logical conclusion, the cashless society could be at hand.

The reality is, however, that they won't – not for a long time anyway. Unpredictable little things like people's preferences will get in the way of this relentless advance. For instance, cash will always be attractive because transactions involving it go unrecorded. The same is not true of cheques, and even less so of computerised banking terminals where every transfer of cash is logged as a lot of 0s and 1s in a solid-state memory somewhere. And computerised shopping – where people 'inspect' goods on a TV screen and pay for them by pressing a few buttons on a terminal linked to a central bank – will not be popular if people think it means they won't get what they want. Although mail-order purchasing, which could be termed one step away from the conventional sales pattern of customer and salesperson in direct contact, has made astonishing advances in popularity, a lot of people won't go near it because they can't see and feel what they are purchasing before they've handed over the cash.

It's therefore stupid to get too carried away with ideas about future 'cashless societies'. However, computer advances will make their presence felt. Already people are reasonably used to communicating with a computer about money – this is what happens when you instruct an automatic cash dispenser to give you banknotes, the computer seeing to it that the cash is deducted from the relevant account. In the US and parts of Europe, shoppers in big stores think nothing of paying for their goods by instructing a computer to amend their accounts in a similar fashion, either directly using banking terminals, or indirectly with a cash-card. Such changes will come about wherever people find they add convenience, and that means fairly often. By the end of the century, most shops in the Western world will have given all their products computer codes (this process is called article numbering) so computers linked to store checkouts can continually keep track of them to help in stock control. In the US, article numbering is already widespread and it is just starting up in Britain.

Other information services, for instance airline ticketing, hotel booking, insurance, travel agencies, theatre booking or railway timetable checking, will all take on board the concept of computers as they find that computer terminals help them to get information and make sense of it, so providing a better service to the customer. Professions which rely not only on advising or otherwise helping clients but also getting information for them from other sources – in this category are lawyers, doctors, architects, surveyors plus less established vocations such as rat catchers and decorators – will also find computers useful.

There is one other broad category of business life that can be considered a kind of service industry. This is the entertainment industry. Dancers, singers, actors, comics will all go on doing the same things very much as before – indeed there should be more call on their services with the extra leisure time we can expect in the future (more of this in later chapters). However, professional sport, which is one form of entertainment, could well be affected. In the US, trainers employ computers to help to work out the best way of training top athletes. College football teams in America also calculate games' strategies using the information-processing capabilities of the machines. Now the idea is spreading to Britain. Various people have propounded ways of raising the standards of soccer players by analysing the pattern of matches with a computer and somehow instilling the results into the minds of footballers in prolonged tactical sessions. Allen Wade, the director of coaching of the Football Association, the world's oldest soccer body, thinks this is a good idea and that computers are bound to make their impact in the sport sooner or later. When people do their shopping with computers, work in microprocessor-run factories, handle all their paperwork with the machines – and then go to watch footballers play as a computer has told them to – we will know computers have well and truly arrived.

Chapter 8: Greater knowledge, more healthy lives?

In the spring of 1972, Britain's Education Secretary gave her seal of approval to an important five-year experiment to test whether teaching with computers was feasible. The Minister's name was Margaret Thatcher. If her fortunes have risen in the meantime, it must be depressing for the advocates of computer-assisted learning (normally abbreviated to CAL) that theirs have largely stood still. Learning with computers is a long way from being established, both in Britain and elsewhere.

The big question is: will cheap microelectronics make a difference? The answer is a lot less clear-cut than in the case of the other areas of application we have discussed in the last few chapters. Likewise it is impossible to say with much certainty what effect the new technology will have in other social services such as health care and welfare. Microprocessors are not necessarily going to make us better schooled and better cared for. Anyone who says otherwise is either an electronics salesman or wants his head examining. The reason is that the software in these application areas is very difficult to produce. The software must be such that it makes the machine interact with the user on a close personal level that is not always required for some of the other uses of computers we have covered. A lot of time, expertise and money is required to write difficult software. And unfortunately these are things that are often in short supply in the public services, particularly in financially-squeezed Britain, that are responsible for education and welfare.

All this does not mean that computer applications in these areas lack promise. For instance, the commercial ambition that led Texas Instruments to sell its 'Little Professor' toy, a gadget slightly bigger than a calculator that asks the user to key-in answers to simple arithmetic problems, has probably done a lot to help children learn the basics of addition and subtraction. It is a fact of life that both computer-assisted learning and computerised gadgets in health care can be expected to progress the furthest where there are companies involved whose interest in these subjects is primarily that they can make money out of them. It must be annoying for ardent CAL backers that Texas Instruments' mass-produced toy has done more to advance their cause than the official Thatcher-launched CAL programme.

This project, which cost £5 million (half provided by the UK government) sponsored thirty-five CAL courses in schools, universities, industry and the armed forces. It covered subjects as far removed as electronics, remedial reading, medical diagnosis, history and telephone maintenance. Richard Hooper, the project's director, summed up the concept's usefulness in his final report: 'The major benefits to come from investment in computer-assisted learning are likely to be qualitative in nature – teaching new topics that could not be taught with existing methods, assisting in the teaching of "higher order" skills such as decision-making and problem-solving, and improving the quality of existing teaching by tailoring it more sensitively to the individual needs of the active learner.'

Hooper's main aim was to lay the ground for CAL by making it available in a significant number of institutions in Britain at reasonable cost. This happened. But since 1977, when the programme ended, CAL has failed to take off in the way that Hooper must have hoped for.

What exactly is CAL? It is the interaction between a person at a computer terminal and the resident software in the machine in such a way that the person learns something useful. This is a pretty hazy definition, so let's give some examples. A person playing with the keypad on Texas Instruments' 'Little Professor' (which can be thought of as a rudimentary terminal) is partaking in a form of CAL. The little screen on the gadget displays a question, for instance '3 × 2 = ?' and the user has to key-in the answer. After a number of questions and answers, the device gives the user a score to tell him how well he has done.

This is a simple version of the 'fact-drilling' type of CAL, one of the four broad ways in which someone can learn with a computer. When it is fact-drilling, the teaching software instructs the student in a parrot-fashion method similar to that by which most children learn their multiplication tables. (This is assuming they have to bother any more, now we live in a world populated by hand-held calculators.) The computer gives the user questions, and marks the answers right or wrong – if all goes to plan the person learns something as he goes along.

Other types of CAL are more sophisticated. The software either reveals something; or acts as a simulator or a time-saver. In the first, the computer gives the student an insight he would not otherwise have had. For instance, in a question-and-answer session conducted via the keyboard and screen, the computer might open the student's eyes to a muddle he is in over a particular area of work. Thus the computer might reveal to a student who is a managing director of a company that he has difficulty making decisions or, if it is a particularly helpful kind of computer, it might go further and offer the advice that the man should resign.

In the second case, where the computer is a simulator, it will draw graphics on the screen to make the student feel he is somewhere other than sitting at the terminal, for instance at the controls of a car or aeroplane learning to handle it, or (for chemistry tuition) in a chemistry laboratory doing an experiment. In the final type of CAL, the computer saves time when, as part of the learning program, the machine works out a routine mathematical equation to save the student having to do it, or routes him to a particular point in the program so he doesn't waste time doing something too simple.

A good example of a CAL program is one devised at the

Massachusetts Institute of Technology called Movie Map. In this, a computer helps someone unacquainted with a particular town to get to know it through films and slides that the computer arranges to be flashed up on to a screen. While sitting in his armchair, the student can then imagine he is walking along roads through the town according to a set route that he has fed to the computer. He can even 'look inside' buildings that he passes by courtesy of displays of internal layouts. And at some places the user can interrogate the computer to ask for more details about some of the buildings he meets on his computer journeys – for instance he can demand to know the menu in a restaurant that the route takes him past.

Other kinds of CAL software can take the student through a series of decision-making steps to learn about, for instance, the best way to run a business. At the Highbury College of Education in Portsmouth, the catering department employs the machines to instruct trainee chefs. The student keys-in answers to questions that the computer poses about, say, the quantity of food needed for a certain number of guests in an imaginary hotel, or how much money that institution should set aside each week for meat and vegetables. The computer can even help the catering student work out details of menus. Another learning programme, devised at the National Computing Centre in Manchester, is aimed at engineers who want to know about the latest concepts in microelectronics. Once again the computer takes the student through a lesson by asking him to draw lines on the screen, answer questions via the keyboard on his terminal and so on.

The beauty of CAL is that it can be flexible. Interaction need not be limited to the user feeding in information by his key buttons and the computer responding via text flashed up on the screen. For instance, in another project at the Massachusetts Institute of Technology called Dataland, the user sits in a chair and controls the computer with a joystick, rather like the ones found in aeroplanes, and touch-sensitive pads. By moving the stick in different directions, the student instructs the computer to display a range of graphical information – for example maps of towns, or pictures of individuals in whom the student is interested.

Probably the world's most advanced CAL system is called PLATO. The acronym stands for something or other that we need not bother ourselves with, although undoubtedly the system's creators chose it because it makes people think about the wisdom associated with Ancient Greece. The American computer firm Control Data developed PLATO from work done at the University of Illinois in the early 1960s. Almost all the world's 4000 or so PLATO terminals are in the US. Control Data hires them out at a set rate that works out at something like £10 for every hour the terminal is in use. Each terminal can be connected via the telephone system to banks of teaching programmes lodged in central computers either in North America or Europe. (In this, the link between terminal and computer is rather like the connection between a viewdata set and a central viewdata computer.)

Participating in a PLATO teaching programme is a delightful experience. The graphics quality is very good; the response time between different parts of the programme (for instance, between one question and the next) is short so that the user doesn't have to twiddle his thumbs waiting for the computer to respond; and the lessons are imaginatively thought out. For instance, the computer might instruct the user about what some people might think are dull subjects – such as mathematics and spelling – with the help of little illustrations of animals that make the lesson far more interesting. A lot of children would find PLATO fun, never mind whether they learn anything from it. Control Data has certainly not made much money out of PLATO: it appears to run the system as an entertaining sideline to the more profitable sides of its business such as selling big computers.

The CAL that is available today certainly shows promise. But is it effective? And how much does it cost? To answer the second question first, it costs a lot. This is one reason the concept has failed to catch on in a big way. Admittedly the decreased costs of computerware mean that an intelligent CAL terminal (one that contains some computing power of its own) in 1979 cost about £1000, a third of the price three years before. This results in more people who can afford to buy a CAL terminal.

Yet the biggest cost of CAL is in the software which, translated into everyday terms, means the time and effort that goes into writing it. A one-hour teaching program can take hundreds of hours to prepare. If the program can be used many times, this reduces the expense because the costs can be shared out over a lot of users. But in the absence of a concerted effort to establish CAL, the scope for the transferability of teaching material in this way is likely to be limited.

Britain's national CAL programme put the costs of computer-learning at £5 to £15 per student hour. This is a lot more than it costs to pay teachers: the CAL project put this at 60p to £2.50 per student hour. So there are definitely no economies to be gained through CAL supplanting people – even if this were desirable. (Normally it isn't: the CAL project team made the point that CAL is a useful aid to conventional teaching not a way of abolishing it.) An average CAL project can cost £25,000 per year to run, plus £25,000 start-up costs, figures that will make many education authorities immediately turn off.

The question about the effectiveness of CAL has to be answered cautiously. It depends on what is being taught. Special subjects, for instance those in which people have to deal with a computer anyway, can be effectively taught with CAL. Thus British Airways every year teaches with CAL about 1000 of its staff to handle the computers that administer seat reservations on airline flights. There are about fifty other companies in the UK that use CAL as a routine part of training – these include Tesco, ICL, NCR and Barclays Bank. On CAL's use in training, Richard Hooper sums up: 'The application of CAL to industry is a straightforward, pragmatic thing. As more of the work that companies do relates to computers, it will be inevitable that people use computers in training. In non-computer related tasks, however, the field is wide open. You're not going to teach people brick-laying with a computer, but you can tackle those tasks that relate to problem-solving and verbal skills.'

Other recent official comments on CAL are not all that enthusiastic. In a 1978 report on an American CAL system called TICCITT, the US Educational Testing Service said: 'Computer-assisted instruction by itself guarantees no magical solutions to the problems faced in education. Results will depend on the instructional nature of the curriculum materials, on the manner of implementation, on the nature of the student population and on the role played by teachers.' The UK's Department of Education and Science said in a paper in March 1979: 'CAL has not so far established itself widely in schools and in non-advanced further education and there may still prove limits to its usefulness despite the availability of new and cheaper computers.' It is estimated that by 1979 about a quarter of Britain's 6000 secondary schools and virtually all the 80 or so universities had experience in CAL. But the numbers of students participating has certainly not grown much (if at all) beyond the 20,000 who experienced CAL in the final year of Britain's national project in 1976–7.

To sum up: there will be no mass changeover from teachers to machines. Children will not come home from school complaining that their maths computer is a nasty old so-and-so because it has made some tetchy remarks on its print-out. However, CAL will gradually become more important in three main areas. One is in those circumstances where using a computer is a vital part of the job to be learned – as in the example of the airline staff handling ticket reservations. Second, CAL will be a useful way to teach subjects which by their nature would require a brilliant instructor indeed to better a computer with a reasonably good program. Thus a simulator that teaches people to fly aeroplanes (such equipment has been around for years as it happens) will never be replaced by a person standing in a classroom with a blackboard and a piece of chalk trying to show how to control a plane.

Lastly, and this is most important, CAL will have a bigger role to play as people have better access to computers. As information networks such as Prestel grow, they could have a lot of potential as a teaching medium. In a decade or so, instead of sitting down at home to watch TV, people could switch on their Viewdata set and select a particular subject they want to learn about. The relevant programme would flash as a series of digital signals along the telephone wires, from the central computer in which it spends most of its time, to lodge in the householder's set. The lessons need not be particularly academic. They could range all the way from needlecraft and keep-fit (providing the person has got a big enough living-room in which to move around) to car mechanics. (In the latter, the student would be better off with a Viewdata set in his garage – but that need not be a problem.) People could take half an hour or so of tuition at a time, so they could fit the lessons into their schedule as they wish, without having to go to evening classes at appointed hours. The Open University at Milton Keynes has spotted the potential of using computers to help to teach its largely home-based students. The University is experimenting with a pilot network of Prestel-compatible Viewdata sets which eventually could supplement the teaching material it supplies its students via TV, radio and books.

Still on the subject of accessibility, the real breakthrough will come when computers are so much a part of life that school children will be given calculator-type instruments to 'go home and play with' and which, almost incidentally, help them learn subjects on the curriculum. The gadgets would be a vital accessory to conventional school work. There is no reason why engineers should not be able to make gadgets rather like Texas Instruments' 'Little Professor' that teach children about, for example, history and geography. In time, such devices could have sophisticated software programmed into them, displays the size of a small TV screen and good graphics capabilities. The latter would permit the display of quite complicated diagrams – the layout of the Battle of Agincourt for history, the design of an internal combustion engine for physics and a map of the Pyrenees for geography.

If costs and availability are the keys to the usefulness of computers in education, then much the same is true in health and medicine. There are two main ways in which computers play a part here. The first is in the expensive medical equipment found in places like hospitals and clinics. For years computers and other electronic gadgetry have infiltrated into this kind of apparatus. The computerised part of a piece of medical equipment will either act as a straightforward data logger, for instance in collating tables of blood pressure readings and working out their significance, or it will control the operation of diagnostic apparatus such as X-ray scanners. According to the US magazine *Electronics*, the world medical electronics market in 1979 was worth $3.77 billion, split up according to Table 3.

TABLE 3: World sales of electronic medical equipment 1979

	$ million
X-ray equipment	1727
diagnostic	811
therapeutic	160
prosthetic	709
surgical support	68
other	301

The second kind of computer-assisted medical system is what we can call the personalised health machine. This is the medical equivalent of the calculator or the home computer. Such instruments are often useful in helping disabled people. An example goes under the name MAVIS, which stands for Microprocessor-based Audio Visual Information System. This is a terminal that is a severely-handicapped person can operate with his teeth or with a small joystick, to call up information (in the same way as able-bodied people read books), receive messages and write letters. Engineers at Britain's National Physical Laboratory build MAVIS as well as several other devices aimed for medical use including two called Muppet and Mickie – the inventors clearly liked giving their creations human names.

Muppet is a talking calculator that can be programmed to do a variety of tasks. According to one of its creators, Dr Chris Evans (who died in 1979 when he still had a lot more to offer the world of electronics), mentally-handicapped people could be given Muppets as 'talking memories' to help them keep track of events and ideas that otherwise they would not remember. Mickie is a box the size of a small TV that doctors can use to help them extract information from patients. Rather like a sophisticated CAL programme, Mickie's resident software asks the patient a series of questions, interspersed with chatty remarks such as 'How are you today?' which are supposed to make the person feel he is getting on well with the computer. According to NPL's workers, the strategy works so well that many people like chatting away to Mickie better than holding a conversation with their doctor.

Mickie is in pilot production and costs around £2000. A few doctors around Britain are trying it out. The idea of Mickie is that it saves a doctor's time by presenting him with a print-out of information about his patient's condition without him having to spend a long time laboriously questioning the patient to find it out. Instead the doctor leaves the person in a room alone with the machine and when the interrogation is complete gets on with the job of examining and treating him.

Blind people are good candidates for computerised help. Telesensory Systems of Palo Alto, California, a pioneer in developing electronic systems for the blind, sells several interesting devices. One called the Optacon translates written text into Braille; an add-on device incorporates a synthesiser chip so that the text is spoken out loud. A second company,

Kurzweil Computer Products of Cambridge, Massachusetts sells a similar machine. Both types of speaking 'reading machines' cost about £10,000. Telesensory makes another machine that costs about £1000 and records Braille characters in a similar way to sounds on a cassette tape. Microprocessors can also be employed to help the deaf or people with speech problems. Telesensory sells a £2000 device with which a handicapped person selects words, phrases or symbols that he wants to organise into a sentence by pointing to them on a board; the finished sentence is then displayed on a small screen. A British company based in Surrey called Medelec sells a similar system.

These gadgets are all very expensive and beyond the reach of most disabled people, who almost by definition have less money to spend on electronic devices than most people, unless they are helped by charities or government funds. Possibly the best kind of help the disabled could get from microelectronics is the development of home information systems such as Viewdata which could let them keep in touch with events, communicate with acquaintances and summon help – all at the flick of a few buttons.

Conceivably, these kinds of developments could help everyone enjoy better health. For instance, medical aid could be available through Prestel, even some form of diagnosis to add to raw information about what to do in an emergency. Face-to-face interviews with remote doctors over new communications links cannot be ruled out. And if you accept useful information on how to take care of yourself, in other words exhortations not to do the things (smoking for instance) that cause ailments which then have to be cured, then data networks could help just by making it possible for people to get access to advice. Whether or not they would take it is, of course, a different matter.

Chapter 9: Death agents

Technical advances do not just happen by themselves. They require some kind of spur to trigger them off. If we look back to the 1930s and 1940s, it is clear that the special demands imposed by the Second World War prompted the development of the first computers. And so it came to pass that, after this war ended, the requirements of the next generation of military men had a lot to do with the emergence of the technology of microelectronics.

The end of the 1939–45 war marked the start of a new period of discord among nations. The world-wide state of uneasiness caused a number of military flare-ups, mainly fairly small ones although sometimes, as in Korea and Vietnam, they threatened to start a more serious, global conflict. There was now, of course, a new dimension to war-mongering. The world's superpowers, the US and the USSR, and their allies, tried to keep their aggression latent for fear that once either side unleashed it, a nuclear war would result in which there could be no winners. Outstanding scientific advances in physics led to the nuclear bomb. Indeed were it not for the awfulness of their invention, the scientists who made the bomb would be universally revered for their brilliance. However, what makes effective nuclear weapons, as well as many other elements in modern warfare, possible today are steps forward in the branch of science in which we are more interested – electronics. Without sophisticated electronic control systems, present-day bombs and missiles would be about as useful as a runaway train careering out of control down a slope. Small-scale, reliable electronics adds direction to tag alongside the power.

The world spends about £190 billion annually on its military forces, with the NATO and Warsaw Pact countries accounting for seventy per cent of this. The US and the USSR are responsible for more than half the total spending. The proportion of this amount of money that goes on electronic systems is virtually impossible to judge. But such systems are a vitally important part of both the weaponry and the detection and information equipment that is deployed before the weapon is launched to find out where it should be aimed. Control and accuracy of weapon aim are vital factors in today's military circles in which strategists are preparing for a war that might last only a matter of minutes. Whichever side has the upper hand in these areas is likely to emerge the winner.

Modern global warfare (or, to reiterate, planning for one in the absence for some years of a confrontation on this scale) can be looked upon in a similar way to a surgical operation. The physician must first explore his patient's condition, establishing with accuracy what is wrong and where to make the incision. It is only afterwards that he plunges in with his knife. In warfare, too, the analysis comes first, relying on up-to-date information on the enemy's whereabouts, his strengths and his likely moves. Next comes the strike, which must be quick, accurate and deadly. Computers help enormously in both stages of the war-mongering operation.

The world was not slow to realise this. During the Second World War, the first fledgling computers were called upon in a limited way to help in the analysis stage. They were useful in processing information from radar, working out ballistics calculations and breaking enemy codes. It will be in the Third World War (if it comes) that the computers' day will have arrived in the second, strike, stage of military conflict. Today's missiles are so loaded with computer power that they can hit their targets with an accuracy of ten to a hundred metres, even when launched the width of a continent or an ocean away. Ground-based artillery employ lasers and sophisticated electronics to reach their objectives with similar precision. Microelectronic advances have spawned a new industry – avionics, or the provision of electronic systems in aircraft. Such equipment in military aeroplanes increases their usefulness to a huge degree and covers navigation, identification of enemy targets, weapon-aiming and flight management.

We have to look at the military's requirements for controlling their hardware (both analysis and strike equipment) to trace the very beginnings of microelectronics. The US Department of Defense, together with NASA, was the godfather of this technology. These institutions sponsored much of the rapid development in electronics that took place in the US during the 1950s and 1960s. The goal of the Department of Defense was to create electronic warfare systems that would beat the Vietnamese in the jungles of Indochina; that of NASA was to make the intelligent controls that would put a man on the Moon. One goal was reached; the other was not: either way, microchips came out of the electronics race. The US government's involvement in the technology reached a peak in 1960 when it purchased, either through the DoD or NASA, approximately half of all the circuits that American firms produced.

As well as making possible electronic equipment that is small and relatively cheap, densely-integrated chips offer two characteristics of even greater importance to military users – lightness and low power consumption. (Actually, electronic circuits for the military are usually not as cheap as they would be for civilian applications. This is because the circuits have to be more reliable and more rugged to meet the rough handling they are likely to encounter.) Lightness and low power usage are crucial where the equipment either has to be portable, for instance where it has to be carried by infantrymen on the battlefield, or where it must form part of a payload in an aircraft or missile where the weight is tightly limited by factors such as fuel-carrying capacity. Because they are interested in these kinds of applications, the military saw the evolution of integrated circuits as a godsend. In the US, the armed forces encouraged this evolution with a series of big contracts to private industry where the technical work mainly took place.

If, by comparison, the UK largely stood by and watched it happen (it did not, after all, have a Vietnam war to fight nor was it preoccupied with getting into space) then it can have the satisfaction of knowing that one of Britain's sons, a certain G. W. A. Dummer, will go down in history as the first person to think of the idea of the integrated circuit. An employee at the Royal Radar Establishment at Malvern, Dummer gave a speech in Washington DC in 1952 in which he spoke of the next development after the transistor. He said it would be 'electronic equipment in a solid block with no connecting wires. The block may consist of layers of insulating, conducting, rectifying and amplifying materials, the electrical functions being connected directly by cutting out areas of the various layers.' Dummer's summary adds up to a reasonable description of one of today's chips. Dummer was a military man himself. Quite clearly he could see the advantages of what he described to military applications. But the authorities in Britain took no notice of his ideas, and the initiative for putting them into practice passed quickly to the United States.

The emphasis in military circles today is to speed up the operation of integrated circuits. Most general-purpose present-day devices react to signals in the order of one to ten nanoseconds (billionths of a second). You might think this is quite fast enough. But, surprising as it may seem, there are many applications in the military arena which require switching speeds of the order of picoseconds (thousand-billionths of a second). Such speeds are therefore up to 1000 times quicker than those of today's devices. These kinds of quick reactions are needed, for instance, when a computer sorts out the meaning of rapidly-fluctuating signals such as those emanating from a radar system. It is important for the computer to work extraordinarily fast if the operaotr is to know the results of the processing operation very soon after the signals arrive. In a world war, the fact that an operator gets this information in a matter of seconds instead of minutes could make all the difference between triggering off an atomic bomb, and deciding that the situation isn't serious enough to warrant one after all.

There are several methods of speeding up today's circuits. The most obvious way is for engineers to shrink the distances between components in circuits using new photolithographic techniques. This will have an effect because, generally speaking, the smaller the distance between the components in a circuit, the quicker it operates – this follows as the electrons and holes have smaller distances to travel to do their jobs.

On the horizon, however, is a new technology that takes us into the realms of what may be called Disneyland science – fantastic ideas that are not yet realisable, but could be one day. The technology concerns circuits called Josephson junctions. Made from metallic substances deposited on silicon in a similar way to today's chips, these circuits work only at extremely low temperatures, near to what is called absolute zero (–273°C). At absolute zero, the to-and-fro movement of atoms and molecules that is a fundamental part of every piece of matter is no more. The temperature is so cold that atoms stop. It is as though the mechanism that is at the heart of life seizes up. At this temperature, there is no resistance to the flow of current through materials and they become superconducting. The principle can be used in designing computer circuits that react to current very fast, in one to ten picoseconds: the only problem is that the circuits have to be made extremely cold, but this can be done by immersing them in liquid helium at –269°C.

Virtually the only organisation to seriously study the possibilities of computers built with Josephson junctions is IBM, the world's biggest computer firm. It spends roughly $6 million annually on this area of research (out of a total research budget of more than $1 billion). IBM has pursued this work since 1967. By the late 1980s the company hopes it will have got to the point where Josephson computers are feasible. These machines, which would certainly be valuable to the armed forces, would digest information at the rate of 250 million instructions per second. By contrast, IBM's fastest existing machine in 1979 could manage a mere five million instructions per second.

To try to coordinate, and spur on, the work in private industry geared towards faster circuits, the US Department of Defense started in 1978 what it called the very high-speed integrated circuit programme. This should last for six years and push $200 million into the industry to produce fast circuits for military applications. Such devices would be very useful in the next generation of missiles. The circuits would be packed into the missiles' warheads to compute with great precision, from data fed to it by radio waves, where it should be heading. If engineers can invent the computerised typewriter, or the intelligent TV set, there is, after all, no reason why they should not create the intelligent missile. This is a device which, once launched, irrecoverably steers towards its prey independently of any instructions from the ground station. If intelligent missiles do not exist already, then there is little doubt that they will do so before long. And once again the world will have to thank – or curse, depending on whether you are on the side that launches the missile – the technology of microelectronics for making them possible.

The new generation which is just emerging of highly-accurate warheads adds a dangerous dimension to warfare. This is the possibility that one country can deploy its weapons with such accuracy that in a matter of minutes they wipe out all their adversary's missile sites, ensuring that the opponent's weapons never get off the ground. Even more alarming for the prospects of world peace is that the super-intelligent missiles could follow the trail of their adversary's warheads after they had been launched, refusing to be shaken off until the rival missiles collide and destroy each other. The upshot of this is that one country can effectively neutralise all of a second nation's warheads: as soon as any of the latter are launched, the first country sends off its intelligent 'hunter' missiles to blow them up. The result? The nation with the best computerised missiles wins the war – game, set and match.

Much depends on the combatants in this sort of encounter having correct information about the sites of their foe's missile stations and so on. Without these data, no amount of fancy microelectronics-aided missile systems are going to help. This is where we come back to the analysis, or information, stage of this two-stage, 'probe-then-strike' nature of the war business. If there is a Third World War, the people who fight it will be the best-informed military men since wars began.

One of the main reasons for this is the extent of satellite surveillance that now takes place. Many of the satellites launched under the guise of communications or mapping projects are taking 'space shots' of enemy territory for military applications. There is no doubt that many of the pictures from the Landsat series of satellites launched in the US find their way to the American armed forces. The information gives them a valuable indication of troop build-ups, missile sites and so on, in many parts of the world.

On a down-to-earth level, microprocessor-controlled equipment that turns the soldier on the battlefield into the equivalent of a walking telex operator is already available. He has his own back-pack receiver-transmitter with which he sends binary-coded messages either to his squadron's base or to other soldiers. The importance of this type of transmission is twofold. First, binary radio messages – electrical pulses that symbolise either 0s or 1s relayed over the air waves – are much more difficult for an enemy to intercept than the analogue signals of voice waveforms because they are more easily 'scrambled' into an unintelligible code. Second, the transmission of messages in written, not spoken, form means the person on the receiving end has a visible string of words to refer to. This is as opposed to a fleeting set of instructions spoken into his headphones which, in the heat of battle, can easily be forgotten or misunderstood. The words in the telex-type system would normally be displayed on a little screen on the soldier's back-pack. If he wants to refer to them again, they are stored in solid-state memory inside the pack, from which they can be retrieved at the touch of a few buttons.

Another important part of the communications system that lets soldiers both keep in touch with each other, and direct vital information to headquarters about the course of a battle, concerns the connection of the individual squadron bases. In a land battle, for instance, one between the forces of the Warsaw Pact nations and NATO in the plains of central Europe, the squadrons would have to be extremely mobile, constantly switching from one headquarters site to another. Communications links between these sites, each perhaps controlling 2000 or so infantrymen, would be absolutely vital. In the old days, these links would be by simple radio communications: one command post would relay instructions and information to another by spoken messages. However, there is a limit on the volume of words that the human brain can assimilate. Computers do the job of transmitting and making sense of information much better.

The British Army has what it thinks is the most sophisticated computerised message system possessed by any force of infantry for just this purpose. Called Wavell and developed by the British electronics company Plessey at a cost of £6 million (two-thirds of which was software costs), the sytem links up battalions covering the 40,000 soldiers in the UK stations in West Germany. The idea is that the 'telex operator' soldiers feed their information to the nearest base with their back-pack equipment. The base then makes this information available to all the other army outposts that are interested, by means of computer communications equipment of the type that is becoming familiar in people's offices.

Each base has a processor which is rugged and portable enough to be manhandled over rough terrain and can be put up in a matter of hours. The processor feeds into several work stations, each comprising a keyboard, printer and display screen. Soldiers using the system feed into it information in much the same way as they would fill in a complicated form on a piece of paper. The difference is that they punch in the words and characters on a keyboard with the text displayed on the screen. All the messages are transmitted by binary radio links to similar screens in the other army posts. In this way, soldiers can send data about the position and strength of the enemy, in which direction they are heading, the location of friendly battalions, special orders about moving camp for instance, or the positions of any nuclear bombs that are due for detonation.

According to the soldiers who operate the system, it means they can handle masses of information – much of it not very relevant to their requirements – and select the most important bits without more-or-less working themselves into a nervous breakdown. The soldiers say that Wavell cuts the amount of time they are on the telephone to other units taking or sending verbal messages by fifty per cent at least. The result is they have more time to use their brains in other ways, doing things like thinking about general battle strategy and so on. The system is one more indication of how computers can help solve the 'information explosion' that affects many areas of life, the military being no exception.

Plessey earns at least one-fifth of its £650 million or so annual turnover from military contracts. In another project for the British Army, the company has produced a sophisticated telephone exchange system called Ptarmigan. This works all-electronically and is a precursor to System X, although on a smaller scale. Eventually it will be employed in linking up the individual base camps that will use Wavell.

The example of Plessey shows how important contracts for the Ministry of Defence are for many British electronics companies. Ferranti relies heavily on military work for its commercial well-being. The company produced one of the very few European-made microprocessors, called the F100-L, specifically to meet Ministry of Defence requirements. One of the first applications of this microchip will be in an engine-control system for heavy army vehicles such as tanks. In late 1979, Ferranti announced a new £15 million research and development unit in Cwmbran, South Wales, to design microchip products for defence applications. GEC, the biggest UK electronics company with an annual turnover of about £2.5 billion (this includes sales from other interests such as electrical engineering), also does a lot of work which leads to military products. For instance, the Marconi group of companies which GEC owns make control systems for ships, avionic equipment and radars.

Many of the chip-associated technologies which will appear in the commercial market-place in a few years are probably now under development at military research centres. With little doubt one of the best ways to learn about electronics is to get a job at one of these institutions. As examples of this kind of technology transfer from the military to civilian spheres, consider three technical ideas – fibre optics, bubble memories and charge-coupled devices. In fibre optics, light is sent along very narrow strands of glass-fibre the width of a human hair. This constitutes a way of sending signals that is an alternative to telephone cables. Microelectronic chips usually have to be linked to either end of the fibre to do the job of sending the signals and to make sense of them when they reappear. Although fibre optics are beginning to look attractive as a way of replacing telephone cables in public telecommunication systems, the military have done a lot of development in the technology because fibre optics systems are virtually free from interference and use little power. So they are very useful in communication equipment in, for instance, ships and aircraft.

Bubble memories, the second of our three technologies, are chips in which data are stored as charged sections of magnetic material. A charged section (or bubble) indicates the 1 of the computer code; a non-charged section denotes an 0. Once again, bubbles are beginning to appear in the public domain: they offer the advantage over conventional solid-state memories in that they retain information without the power switched on. However, it is interesting that one of the first 'real-life' products in which bubbles appeared is the military communications system Wavell that we have already described. Here the bubbles are useful as storage media because they consume little power and they can be destroyed very easily if the enemy is about to take over a command post, with the prospect of him getting his hands on valuable information stored in Wavell's processors.

Charge-coupled devices are slightly further away from mass application. They are chips made from silicon that react to light in such a way that they 'see' rather like a TV camera. The devices cause a lot of interest in military circles because of their capabilities as sensors. Thus they can be incorporated in the warheads of missiles to take in information about where the projectiles are heading. Charge-coupled devices also have applications in infantry surveillance – for instance they could help a soldier work out the position of snipers when it is so dark that there is not enough visible light around to help him.

These examples show that military leaders often get the bright ideas about how to use technology long before anyone else. The microchip, which the world is now starting to argue about in terms of its possible uses and the changes it will herald in society, is positively old hat to most military boffins. They discovered it long ago in its first and most grisly role as an agent of death.

Part 3

CONSEQUENCES

Chapter 10: Employment – the opportunities

In this chapter and the next we will discuss the effect of cheap computers on employment. Chip-aided machines may not do people's walking for them. But computers could take away jobs.

Changes in technology do not make their presence felt in a vacuum. They are influenced by all the other things that happen in the world. This is what makes forecasting the future difficult. Usually, predicting how technology will change is rather easier than trying to forecast what people will do with the results. For instance, a few far-sighted observers have predicted for years what such things as communications satellites and space probes now make possible. Even if we consider the history of the computer, the development of the microcircuits that comprise today's generation of machines was foreseen in 1952 by the Englishman Dummer. And we saw in Chapter 3 that the prediction in 1964 by Gordon Moore that the complexity of integrated circuits would approximately double every year turned out to be amazingly accurate.

What even the shrewdest crystal-ball gazer finds much more difficult to foresee, however, are events such as wars and revolutions or, on a more mundane level, strikes or other occurrences that affect industrial activity. Likewise, it is almost impossible to predict social changes that may not seem on the surface all that important but whose net effect is large. One such change that has occurred in Britain over the past twenty years is the increased desire of women to go out to work, a social force that has increased the income of the average family but now adds to the country's problems over unemployment.

Another steady change is that more and more people look

upon a decent standard of education as a right. Consequently bigger sections of the population are becoming more highly-skilled and they expect society to gear itself to meet their needs. Probably the most important single event that has happened to the world since 1945 was the Arab-Israeli war of 1973 which led to a quadrupling in oil prices in a matter of months. In the years to follow, the developed world was forced to appreciate that its appetite for energy must moderate. The West was, quite simply, using the Earth's finite stocks of energy faster than was permissible if it wanted to have a chance of a reasonably prosperous future.

How does the advent of the microchip fit in with the impossible-to-predict ebb and flow of human activity? First, let us recap slightly. The first electronic computers invented in the 1940s presented the world with calculating machines that did their work 1000 times more quickly than anything hitherto. But for all this power, the computers were big and clumsy, expensive and liable to go wrong. Since then scientists and engineers have made the components of these computers thousands of times smaller and by a similar order of magnitude much cheaper. Computers also work a lot better than they did thirty years ago – evidence of the truth of the old adage that practice makes perfect. There has been an awful lot of practice in computers since the 1940s.

The result is that from a few thousand in the early 1960s, the world now boasts a few million computers (where a computer is any machine containing logic and memory hardware that is capable of being programmed). By the late 1980s there will be hundreds of millions of the machines. By the 1990s, the numbers will rise to thousands of millions, by which time the devices will begin to outnumber people.

At the risk of being over-simplistic, it is possible to classify all of human activity as the conversion of energy into information, a process which may (or may not) give rise to the creation of wealth. We can summarise this as:

Thus human life depends absolutely on energy, given, either directly or indirectly, from the sun, for all its deeds and misdeeds. Human beings produce information as a fundamental part of life. If you think about this, it is more or less impossible to do anything to change the process. Every act – walking, eating, talking or sleeping – produces a certain amount of information which someone, somewhere, might record (or far more likely, fail to take any notice of whatsoever).

Wealth creation enters the scene on those relatively few occasions where people take the trouble to collect the information, collate it into neat bundles and do something with it. In this way, people took notice of the information from the natural world which let them learn to plough fields or practise rudimentary astronomy to forecast the seasons. In such a way they discovered enough to erect buildings, or, bringing us a little more up to date, to harness steam power in industry. All of scientific and technological progress is due to this bundling together of information – we may call it the great information harvest – as is the creation of businesses, the building of ships and aircraft and so on.

Wealth may come as a result of this information collation. This happens in the building and operation of factories that produce goods for sale at a profit, or of oil rigs or mines which let people pluck valuable natural resources from the earth

Then again, material wealth may have nothing to do with the process, as in the collation of information in the form of literature or music. And of course the endless creation and ordering of information may have entirely destructive results, as you would appreciate if you were on the receiving end of an atomic bomb whose information content, reckoned in the number of man hours of intellectual effort that is needed to create it, is extremely high.

Computers greatly ease many important stages in the processes of information collation, storage and retrieval. They help carry out complex calculations, run machines in factories, and transmit and make sense of huge quantities of data using the telephone system. The evidence for computers' intervention in these areas has been marshalled in Chapters 5 to 9 and I hope that it has convinced you. We can therefore assume that by helping in the vital human activity of information harvesting, computers can play an important role in creating wealth. There could be problems, however. Computers could have such a big impact that they displace large numbers of people from the information-collation process, in other words, take away their jobs; or they could interfere in other ways with previously-established patterns of life.

It is an incontrovertible fact that many of the processes by which people earn their living can be done more efficiently with the help of computers. We learnt this much from some of the arguments presented in earlier chapters. A machine tool that incorporates a computer increases its operator's productivity by a factor of between three and five. Something similar happens if a clerical worker is given a word processor instead of a typewriter. It would follow that, with the advent of computerisation, a nation requires only one-third to one-fifth of the tool operators and typists employed in the pre-microchip age.

Other workers would be similarly hit. Ian Barron and Ray Curnow, in their book *The Future with Microelectronics*, list no less than thirty types of worker who they say computerisation could make redundant. At the risk of being boring, here is the full list: proof readers, library assistants, mail carriers, telegraph operators, draughtsmen, programmers, accountants, financial staff, administrators, secretaries, billing clerks, key punchers, cashiers, filing clerks, meter readers, shipping clerks, TV repairmen, plateprinters, telephone repairmen, light electricians, machinists, mechanics, inspectors, assemblers, operatives, material handlers, warehousemen, sales clerks, stock clerks, and compositors. The list reads like a *Who's Who* of trades and it is easy to argue over which ones should – or should not – be included.

Two factors, however, interfere with the prognosis that productivity increases due to automation will lead to the disappearance of jobs.

The first factor is that the quality of the finished product may be enhanced by computerisation. In this way, new markets can be opened up through sales of goods to people for whom the previous generation of products – made by uncomputerised methods – were not all that attractive. The results of this process are that more people are required by the organisation that produces the items, offsetting, at least partially, losses in employment due to automation. Thus if motor cars can be made more cheaply and with fewer faults with the help of robots, then more people might be persuaded to buy them, boosting the fortunes of the company involved and (possibly) forcing it to employ more people.

The second factor that affects our simple conclusion about unemployment is that the world may actually require more of the goods than the worker produces, especially if they can be made more cheaply. For instance, factory employees may work flat-out with the uncomputerised piece of equipment and still not produce as many items as can be sold or otherwise used. With a new process, a machine operator may be able to turn out three times as many items per shift; but this won't mean he or his co-workers are sacked if his employer can sell this many more.

Office workers are an interesting bunch of people to consider when thinking about how employment changes due to productivity increases can be offset by greater demand. Here we can examine figures assembled by management consultant Ian Mackintosh. For all the technical innovations that have helped increase these workers' output over the past century, the number of clerical staff employed in the UK has increased from 250,000 in 1891 to 3.5 million in 1971 (see Figure 6). Productivity increases due to the development of the typewriter, the telephone and the dictating machine did not mean fewer people were required. Instead, firms' needs for clerical staff mushroomed, presumably because more and more clerical information was needed to help these firms commercially.

The impact of the technical advances was almost completely disguised in the voracious clamour for the information that the clerical workers are paid to create, file and disseminate. And so it may be that the impact of computerised devices such as word processors will go similarly unnoticed by workers in the office of the future. Rather like the ladies in TV soap adverts who are smothered in their baths with lather, we may find that clerical workers in the year 2000 will wallow in the luxury of boundless volumes of information which, for all the power of computerised office machines, need a substantial body of people to produce.

Telephone exchanges provide another example which should give hope to people who believe they are threatened by automation in the form of computers. In the 1920s, when telephones started to become reasonably common in Britain, connections were made by manual exchanges. Operators had to plug wires into sockets to physically set up calls. This worked fairly well for the small numbers of people who had a telephone. But demand has grown so enormously over the past half-century that, had it not been for technological innovation that gradually automated the telephone exchanges to remove the need for manual operation, the situation would by now have reached crisis point. Given the scale of today's telephone

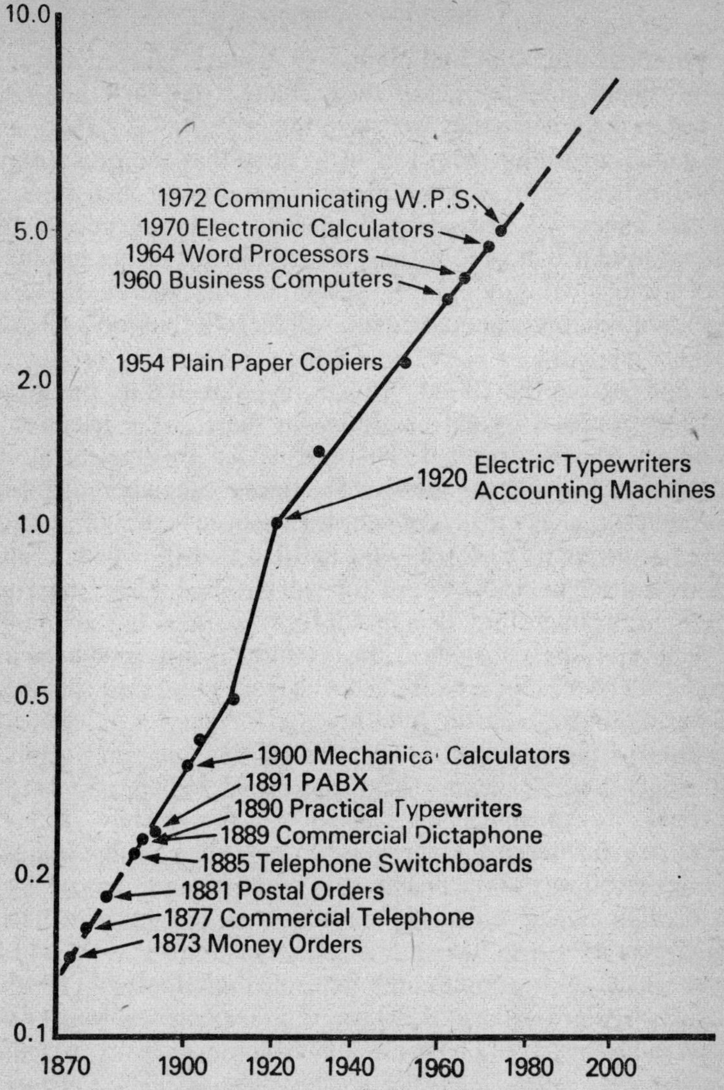

Fig. 6 UK Clerical Workers (Millions)

Source: MacKintosh Consultants

population, if we still had manual exchanges, half the nation would be required to operate the exchanges and the other half to repair them when they went wrong.

It is easy to infer from this little story that the process by which Britain's next generation of computer-controlled System X exchanges will require perhaps half as many people to operate them is no more than a further step in the gradual evolution of the telephone system. Thus there is every likelihood that increased demand will see to it that jobs are not lost but gained.

Figures from the US help prove the point that automation does not necessarily eat into jobs. In the US, the telephone system is much further down the road to an all-electronic, computer-controlled network than anywhere else. Nonetheless, the numbers of people employed in telecommunications increased from 611,000 in 1950 to 810,000 in 1976. However, the composition of the workforce did change. The number of operators decreased from 267,000 to 203,000. Engineers and clerical staff, on the other hand, increased in strength, from 147,000 to 300,000 and from 112,000 to 178,000.

The arguments so far put the case for extra employment owing to enhancement of, or increased demand for, existing products and services. What about *new* activities that computerisation makes possible or desirable? Here, surely, there are further opportunities for new jobs. The obvious examples of ways in which chips can generate employment involve new chip-aided goods that did not appear until microcircuits made them possible. Thus firms that make pocket calculators, home computers and digital watches would not have existed without the chip. And the people who work in these industries would have had to find some other way to earn a living.

Yet probably more significant is the way in which computer-orientated activities can stimulate employment in the industries that spring up to serve these activities. The growth of what are called the service industries is one of the most significant trends of the past thirty years. The shift of people into these areas of employment is the latest in a series of structural changes that have affected the nature of work since life began. As mankind has evolved, technical advances have made many activities easier, thus removing the need for people to carry out physical work.

A simple example is the invention of the wheel. In the post-wheel age, people who hired others to construct buildings could afford to reduce their labour force as the same quantities of building materials could be dragged along by fewer people. Similarly, steam-powered machinery meant people did not have to manually pump water out of the coal mines in eighteenth-century Britain. Children no longer had to climb up chimneys to sweep them once machines did the job. Householders did not have to employ maidservants to pick dust off carpets after Hoover became a household name in the 1930s. What happened to the ancient activity of the human stone-hauler who was made redundant by the wheel? What happened to the eighteenth-century water-pumper, the nine-year-old sweeps, and the 1930s servants? In the main, the answer is simple. They got other jobs.

In much the same way, opportunities exist today for those people whose jobs are displaced by computerisation. The history of the industrialised world since the year 1800 shows several important changes in employment patterns. Due to those, it was inevitable that certain types of people should lose their jobs and move into others. During the spate of factory building that followed the industrial revolution in Britain, people moved out of the so-called primary industries such as farming and fishing to man the new processes. They became workers in what, in economists' jargon, is called secondary industry which includes mining and construction as well as manufacturing.

At the same time, what is called the tertiary sector – service industries – also grew. Most of the new service workers were employed in domestic activities. The new prosperous middle-class that arose from the industrial revolution hired them as butlers and maids. In Britain in 1800, roughly thirty-seven per cent of the working population was in the primary sector, thirty-three per cent worked in secondary industry and thirty per cent were in the service or tertiary sector. Workers in the last two sectors grew steadily until 1900, as those in farming and fishing decreased. The numbers then started to level off slightly until by 1950 workers in the three areas of industry totalled respectively five, forty-nine and forty-six per cent.

Much the same process occurred in other Western countries, although the shift from agriculture-to-manufacturing-to-services usually began a little later than in Britain, a consequence of the industrial revolution starting in the Pennines rather than elsewhere. In 1950 most Western countries had far more of their population engaged in agriculture than in Britain. In France the figure was eleven per cent, in West Germany twenty-four per cent and in the US twenty per cent.

Since 1950 the swing from manufacturing has gained in impetus with the result that more people have gained employment in the service, or non-producing, sector. This trend has been marked by the emergence in the industrialised world of more and more people like local government officers, school teachers, insurance men, barbers and cleaning ladies (see figure 7).

In Britain in 1978, 57.5 per cent of the population was in services, as opposed to forty per cent in 'secondary' trades. Manufacturing industry's share of Britain's 1977 gross domestic product was twenty-nine per cent; seventeen years before it had been thirty-five per cent. The 1970s saw an explosive growth in the numbers of non-manual workers not directly involved in manufacturing. Between 1971 and 1976 there were fourteen per cent more managers and administrators, eighteen per cent more scientists and professional engineers, twenty-two per cent more people in health and social services, twelve per cent more draughtsmen. The total of non-manual workers increased from 10.7 million to 11.8 million or by ten per cent; those in manual trades went down from 13.5 million to 12.8 million, a reduction of five per cent. The figures mark the extent to which Britain has become a society dependent on services.

The creation of jobs in the service sector acts as a sink for those people who cannot, for one reason or another, get the

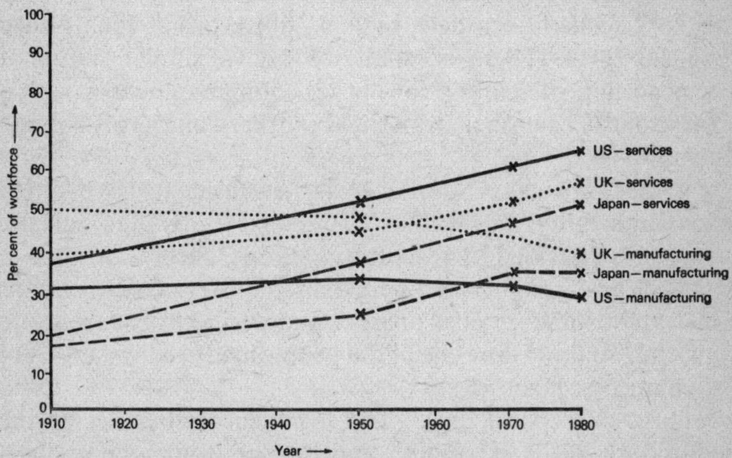

Fig. 7 Percentages of workforce in services and manufacturing for UK, US and Japan.

Source: The Manpower Implications of Microelectronic Technology, Department of Employment

employment they expect in manufacturing industry. Thus wealth caused by the use of computers in 'information-harvesting' work – in offices, factories, construction work – and in the information-processing activities that support these activities leads to a demand for services that are not necessarily related to computers. And this in turn can create employment.

To see how this can work, let us consider a hypothetical engineering firm that employs 200 people, 150 of them machine-tool operators and fifty clerical and administrative staff. The company decides to sell all its existing machinery and tool-up with new equipment, let us say thirty word-processing stations and fifty computerised machine tools. Finding it does not need so many people to produce the same output, the firm sacks fifty machine operators and twenty typists. However, at the same time, the company finds it requires fifteen new computer engineers to supervise the new equipment and maintain it. These the company imports, from outside the immediate area.

The upshot of all this is that there are seventy people looking for work while fifteen new, higher-skilled, professionals have replaced them. There is, shall we say, a similar influx of engineering professionals into other companies nearby with a corresponding number of manual workers and typists made unemployed.

But life is not static. Soon after the new residents have settled into their locality, they want to spend their (fairly high) salaries on services provided by restaurants, shops, barbers, insurance firms, cleaning staff, plumbers and builders. Organisations which offer these services prosper and decide to engage more staff, thus soaking up some of the people made redundant from the engineering factories.

Things don't stop there. The local authorities decide the neighbourhood is expanding and budgets for more teachers and health-care workers, and consequently for the people who will service them – canteen workers, cleaners, office staff. And so the process can go on, one lot of service workers feeding off another. In time, if all works well, the number of service workers should be as great, if not more than, the number of people initially displaced from the manufacturing industries.

This is a microcosm of what has already happened in many developed countries in the past few years. To take a concrete example that concerns computers, the US computer industry employs only about 200,000 people in the reasonably heavily automated work of making the machines. However, the companies that service this industry – a fairly specialised sort of service sector as it deals with things like software and consultancy services – employ a huge two million more. The job creation effect of writing software for computers is considerable – for instance, it is estimated that $1 billion worth of programming has been done for the world's biggest-selling microprocessor, the Intel 8080. In California, only 100,000 people work in the State's chip-making firms. But the wealth this creates produces a ripple effect which leads to some half a million new jobs every year.

A further way computers can help generate jobs, according to one popular thesis, is because they stimulate small-scale enterprise. One person armed with a microcomputer is a formidable businessman. He can be his own accountant, engineer, filing clerk, typist, telephone operator and statistician. In a few years, when developments such as data networks have really taken off to help computer-to-computer communication, and when hardware prices have plummeted further, computerised aids will help the business person even more. Providing people are properly trained to use computers (and this means a big effort in schools) most thirty-year-olds in the year 2000 will have basic computer skills.

At that time, there could be a big incentive for people to start up their own small businesses with the help of computers and only a fairly small workforce. What they do or sell is not really the issue. The story of computers themselves, after all, proves that people can sell almost anything once they perceive that the demand exists.

Politicians all over the world are gradually coming round to realising that to provide the right opportunities for small businesses to prosper is a very good way of creating employment. A study that showed this was conducted by the US Department of Commerce in 1978. This examined six big mature companies (including General Motors), five firms known for their innovation capabilities (including Polaroid and IBM) and five small, young companies (such as Digital Equipment). The study showed that in terms of jobs created per dollar of revenue the young firms had by far the best record. These companies created 410 jobs for every $1 million annual turnover, the innovative firms fifty, and the 'biggies' of the business world only seven.

Supporters of the idea that we are at the start of a new golden age of prosperity, fuelled by the advent of the microcomputer, often point to what are supposed to be long-term cycles in innovation that run through the world's economy. These cycles, known by economists as Kondratiev long waves, comprise a series of peaks and troughs – a peak corresponds to a period of boom high employment and a trough is a recession. Booms appeared around the years 1800,1850, 1900 and 1950. They correspond to the technical advances that produced steam engines; railways; the electricity supply and chemical industries; and the TV and radio (see Figure 8). The next boom, according to this prognosis, is scheduled to reach a peak

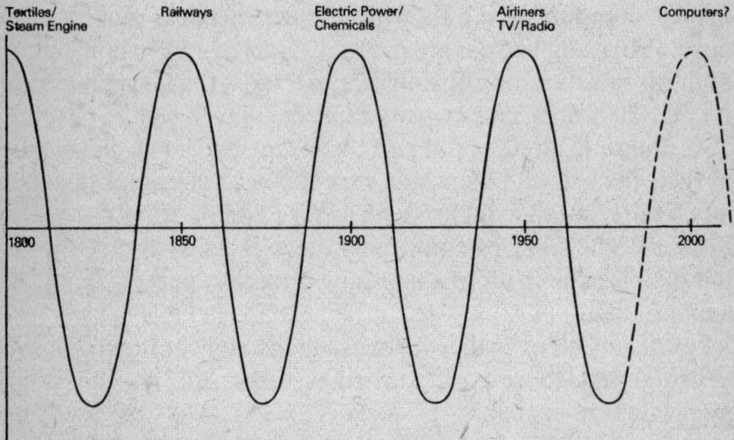

Fig. 8 Kontratiev Waves

around the year 2000 and will be fuelled by innovations in computers and the impetus these give to businesses.

To recap, in this chapter we have considered five ways in which cheap computers can generate jobs. The employment is due, respectively, to enhancement of, or extra demand for, existing products; creation of markets for new chip-aided goods such as calculators; creation of new services in a continuation of a service-dominated economy, and by the shot in the arm that chips can give to businesses of all kinds due to the advent of 'personal computing power'. From an employment viewpoint, this is the optimistic way of looking at the future . . . let us now consider the darker side of what could happen.

Chapter 11: Employment – not enough to go around

The last chapter started from the standpoint that computerisation will involve the loss of jobs in some areas of human activity. However, it went on to present arguments that show how these effects can be mitigated by the provision of jobs in other areas. Some of these jobs, for instance, those in the service or tertiary sector that 'feed' off computer-based activities, could themselves be stimulated by the advent of cheap computers. This chapter will examine just how realistic these arguments are.

The case for cheap computers *not* leading to diminished employment opportunities borrows much of its force from history. Increases in productivity per person due to technical advance have normally been accompanied by more employment, not less. In Britain since 1920, output per worker has, on average, increased at a steady rate of one or two per cent per annum. At the same time the labour force has slowly grown, even though, in the past thirty years, the growth has diminished to just a small fraction of a percentage annually. The big question mark that concerns us as we enter the computer age is whether this long-term trend will continue. After all, why should it? We may, conceivably, have reached a turning point, where the extra productivity brought about by computer-inspired automation, together with other important economic forces that we shall now consider, is leading the world into a new age of high, permanent unemployment.

A lot of employment theory concerns the name of the man who stamped his influence on Western governments in the mid 20th century, John Maynard Keynes. According to economists in the pre-Keynesian era, ups and downs in employment, inasmuch that anyone took any notice of them, were vagaries of the market system that normally sorted themselves out in the end. According to accepted theory, high unemployment would lead to workers agreeing to lower wages for the same amount of output. At this point factory owners and other employers would find they could afford to invest in more capital equipment. They would plan for bigger output and hire more workers to achieve this. When employment rose to the point where it was difficult to recruit more men, wages would rise, and investment and production would slow down. Eventually this would lead to lay-offs in the next recession and high unemployment, at which point the whole process would restart.

The world ran into difficulties in the 1920s and 1930s when unemployment did not 'sort itself out' in the way predicted. Instead, it persisted at an alarmingly high level. Keynes showed that governments could take responsibility for employment and successfully treat it by stimulating demand for goods. This the Western governments did when they adopted Keynesian policies in the post-war era, mainly by boosting public spending, not worrying if the process resulted in a budget deficit. In other words, governments were prepared to spend more than their income to hold down unemployment at a reasonable level.

A second unwelcome by-product of the policy was a moderate amount of annual price increases – what economists call slow inflation – caused by constantly spiralling, government-led, demand. In Britain between 1945 and 1970, this inflation was just a few per cent per year. But it was a small price to pay for warding off the bogey of mass unemployment. Over this period, the proportion of people out of work was kept to one or two per cent – a far cry from the twelve per cent or so of the 1920s.

During the 1970s, however, the Keynesian approach failed to work. The chief difficulty was that increased government demand did not prevent high unemployment. Instead it resulted in unacceptably high inflation. The failure should not really have come as any surprise. Keynes's theory coped with the problems of the 1930s and 1940s: there is no reason why it should have been as successful in dealing with the different problems thrown up several decades later.

A major cause of the inflation in the 1970s was the steep rise in oil prices in 1973. This hit a world that was already on the crest of an inflatory boom caused by heavy deficit-spending by the US to finance the Vietnam war, and by expansionary economic policies in the major countries. This had brought big increases in wage rates in the early 1970s.

The events of 1973 were followed by big increases in the prices of transport and chemicals, and by further price rises through the decade that affected raw materials used by industry. The effect of this large transfer of wealth from the industrialised West to the oil-rich OPEC countries, as well as deliberate policies adopted by governments to combat inflation, was a significant downturn in the economies of the industrialised nations. Growth levels, measured as percentage increases in gross domestic product, were reduced appreciably from those of the 1960s 'boom' era. The slow-down was reinforced by another big rise in the price of oil in 1979. During the year the OPEC nations hoisted the price of a barrel of oil to the $30 mark. At the beginning of the decade it had stood nearer $2. This represented an increase, allowing for inflation, of 500 per cent over the ten years. The face of the 1970s, mostly a depressing one, for the developed world is summed up in Figure 9 and Table 4.

So far in this chapter we have hardly mentioned computers. But their impact, superimposed on the economic trends that already exist, could be considerable as far as unemployment is concerned. From the analysis just presented, we can say that one of the reasons for unemployment is a slow-down in economic activity of the sort that has happened between 1974 and 1980. Further reasons for the Western world's high unemployment figures are (a) increase in labour costs in developed nations by which workers help to price themselves

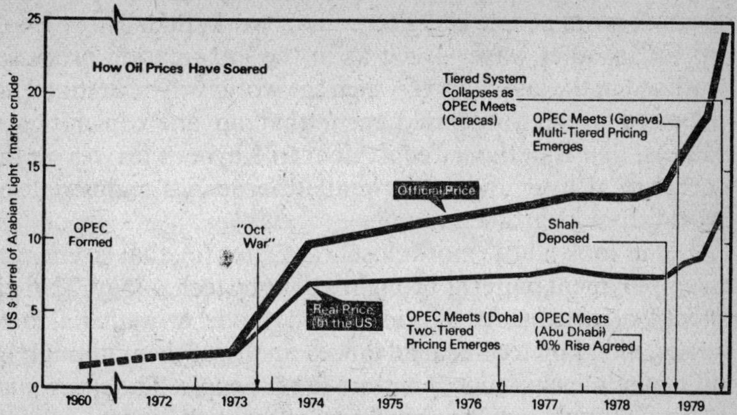

Fig.9 Oil Price Rises 1970-1980

Source: *Financial Times*

TABLE 4: Statistics for economy of selected countries 1963–78.

Sources: *Technical Change and Employment* by Rothwell and Zegfeld; London Business School.

Country	Growth (per cent per year)		Inflation (per cent per annum)		Unemployment per cent of workforce	
	63–69	70–78	63–69	70–78	62–73	74–78
US	6.1	2.6	3.0	6.6	4.9	7.0
Japan	14.0	5.0	5.0	9.6	1.3	1.9
W Germany	5.5	2.9	2.5	5.0	1.3	4.2
France	6.4	3.4	3.6	8.6	1.8	4.0
Italy	7.2	3.1	3.4	12.2	3.6	4.8
UK	3.3	0.9	4.3	12.0	2.4	4.6

out of the market; (b) changes in demand patterns by which, for instance, people no longer want to buy goods produced by high labour-intensive methods; and (c) changes in technology which mean that fewer people can produce the same quantity of goods.

All these factors can be said to represent structural changes. They are changes that Keynesian demand–management tricks can do little about. It follows that if the sum of all four factors leads to more people being put out of work, than can be made up for in other ways – such as in the job creation processes outlined in the last chapter – then the world will be destined for a period of high unemployment that no one can mitigate. Unless, that is, a theory equivalent to Keynes's arrives on the scene to deliver the late twentieth-century's industrialised world from its problems.

Let us look a little more closely at factor (c), that is, changes in employment patterns brought about by technology. There is plenty of evidence that shows how modern industrial processes, aided by technical advances (many of them, though by no means all, chip-aided) require fewer people. The electronics manufacturing industry has been affected the most. Today's pieces of electrical equipment are generally simpler to make than previous generations because of the results of large-scale integration of electronic components.

In many pieces of equipment, moving mechanical parts can be replaced by electronic ones, often highly integrated on to a single chip. Thus an electromechanical telex machine takes seventy-five man-hours to manufacture whereas an electronic machine takes only eighteen man-hours; an electronic taxometer takes 3.7 hours compared with twelve hours for a mechanical one; a mechanical watch takes some 1000 operations to assemble while someone making an electronic watch has to handle just five components. It takes four hours to assemble an electronic calculator instead of twelve hours for a mechanical machine; an electronic typewriter requires half the labour content of an electromechanical model. The manufacture of a telephone or a TV set once required teams of people to assemble the individual components that make up the apparatus: today many of these components are near-invisible and integrated on to a chip. The assembler's job has been done for him. A modern TV set contains about 500 different parts: in the early 1970s the figure was twice that. Similar figures exist for telephone exchanges and other pieces of telecommunications equipment. They are reflected in the numbers of people employed in these (relatively successful) industries whose output is increasing. The UK's telecommunications firms employed 90,000 people in 1973. By 1979 this had slumped to 65,000. There were 70,000 people employed making colour TV sets in Britain in 1973: six years later the figure was 46,000. Scotland, with a fairly strong home electronics industry, employed more than 40,000 people in this area in the early 1970s. By 1978, the figure was down to 34,000, even though output over this period had gone up by more than thirty per cent.

We can look elsewhere in the world for very much the same story. Western Electric, the telephone manufacturing arm of the giant US telecommunications company AT & T, employed 19,000 in 1976, roughly half the total in 1970. The electronics firm NCR gave work to 70,000 people in 1976, 30,000 fewer than six years before. Yet over this period turnover per employee had doubled. Numbers employed in the phenomenally successful US electronics industry, whose dollar output is increasing at the rate of roughly thirteen per cent annually, is only just over a million. In 1969, in the days of labour-intensive processes that involved assembly or valves, resistors and all the other paraphernalia of the pre-chip age, employment stood at 1.25 million.

More than anything else, employment in the electronics industry is shifting towards the technically-qualified worker. This is bad news for the untrained person made redundant by a struggling mechanical engineering firm who turns to the successful electronics industry for a job. It may be booming but not for him. The extent of the change is shown in figures supplied by Siemens, the West German electronics and electrical enginering company. Over the period 1962 to 1977, the firm's workforce stayed roughly constant at around 170,000. In that time semi-skilled shop-floor employees dropped from forty-four per cent of the workforce to thirty-two per cent; likewise, skilled factory workers saw their numbers reduce from twenty-three per cent to twenty-one per cent. By contrast, the proportion of technical staff in the labour force rose from eighteen per cent in 1962 to twenty-nine per cent; and clerical workers were also more in evidence, their numbers increasing from fifteen per cent to eighteen per cent.

We can now look at industries where the reasons for decreases in number of employees are rather more complex. In Britain, there are something like one million fewer job opportunities in the manufacturing sector as a whole compared with the mid-1960s. That not enough of these job opportunities have been replaced in the mushrooming service sector is the reason for unemployment levels that hover around the 2 million mark. Another factor is that the size of the workforce has increased considerably over the past ten years or so mainly as a result of more women going out to work.

But why did unemployment happen in the first place? The answer is a mixture of all four factors mentioned earlier. Technical change has played its part in increasing productivity, and hence reducing the requirement for workers. But more important has been these industries' decline in activity as, for various reasons, demand for their goods declines. This can be either because the world just does not need so much of a certain kind of product, or because consumers prefer to buy from a rival, possibly overseas, company on the grounds of price or quality.

Examples of British industries in decline on account of these two factors are steel-making, whose output has sagged in phase with the slow-down in demand on the part of user industries such as motor cars and construction, and hi-fi equipment which has suffered much from consumers' greater preference for overseas equipment, for instance from Japan. Even when the output of the British Steel Corporation was increasing from 1950 to 1973 (when production almost doubled), the number of employees fell from 630,000 to 470,000. This was mainly due to improved technology. The future, compounded of both technical improvements and decreasing demand, looks bleak for people whose ambition is to work in the steel industry.

Textile machinery is another example of a UK industry in

decline because it cannot sell enough of its goods. In 1954 the industry accounted for thirty per cent of the world's export market: in 1975 the figure was eleven per cent. In that time employment halved to roughly 30,000. Such changes in employment patterns are not confined to Britain. For instance, the US steel industry, whose output over the period 1954 to 1976 was roughly constant, needed 470,000 workers at the end of the period as opposed to 630,000 at the beginning.

Britain, however, seems to be in more trouble than most countries as a result of the shrinking in its manufacturing base. Since 1961 its share of world trade in manufactured goods has fallen from 12.7 per cent to 7 per cent in 1978. Over this period, imports of manufactured goods, expressed as a ratio of imports to gross national product, has increased from 4.6 to 14.2, a sure sign that the country's industries are failing by comparison with those from overseas. The complete picture is given in Table 5.

TABLE 5: Trade record and imports for UK and other countries.

	SHARE OF WORLD TRADE			*RATIO OF VOLUME OF MANUFACTURED IMPORTS TO GNP*		
	1961	*1969*	*1978*	*1961*	*1969*	*1978*
UK	12.7	8.6	7.0	4.6	8.0	14.2
Rest of EEC	33.9	35.0	37.1	6.1	10.1	15.8
US	17.9	15.5	11.5	1.5	3.4	4.5
Japan	5.0	8.3	11.5	1.8	2.2	2.4

Source: Cambridge Economic Policy Review

Further evidence of the inability of British industry to produce the goods that home consumers want is provided by a harsh look at the mini-economic boom in Britain in 1978. Consumer spending increased over an eighteen-month period by some thirteen per cent, sales of cars reached record levels and imports went up by twenty per cent. Much of the money the public spent went on goods and services from abroad. It boosted employment in foreign countries, but hardly at all at home. Few home companies reacted to the surge in consumption by producing more goods and recruiting more staff: unemployment in Britain fell by only about 100,000 over the period.

If we do not consider the advent of microelectronics at all, the future on the unemployment front looks bleak for Britain. Given the UK's chronic record in manufacturing, over the next few years employment in this sector seems set to fall even further. It is difficult to see how the service sector, quickly as it has grown over the past twenty years, can grow fast enough to make up the jobs lost in this area. One of the most gloomy predictions of what could happen came from the Department of Applied Economics at Cambridge University in a forecast in March 1979. The department said that without the introduction of import controls and, possibly, a devaluation of the pound to curb import penetration, there could be almost four million unemployed in Britain by 1990.

Other commentators have made much of the fact that Britain is hardly a world leader in applying microelectronics in industry. A useful pointer to this was a Department of Industry survey in 1978 which showed that only five per cent of British companies used the technology to improve either products or processes. Fifty per cent of the companies in the survey had never even bothered to find out how microelectronic-based techniques – for instance extra automation in factories or word processors in offices – could help them. Some people have concluded that Britain should make an out-and-out-effort to use chip-aided techniques in manufacturing industry to stop this sector disappearing once and for all as an employment-generating force. The argument is that microelectronic techniques may require less people than orthodox industrial processes – but they could be the kiss of life to a previously uncompetitive British company.

Thus a 1979 report on robots from Britain's Advisory

Council for Applied Research and Development stated: 'We are in great danger of being left a long way behind in the application of programmable automation and robotics . . . If these deficiencies are not overcome the United Kingdom will find that its manufacturing is increasingly non-competitive as production becomes highly automated in the more advanced countries or moves to areas of plentiful low-cost labour in the less developed countries. Further, automation effectively applied not only raises productivity but also improves quality. Failure to use these techniques is likely to lead to loss of market share through inferior quality as well as uncompetitiveness on cost. These are greater threats to employment than the displacement of labour by machines.'

Another report in 1979 from a working group in the Department of Employment echoed this theme. Charged with looking at the implications for employment thrown up by microelectronic technology, the group concluded: 'In the secondary sector it seems probable that the overall growth in demand for products as a whole, whether or not affected by new technology, over the next five to ten years will be a more significant determinant of employment levels than technological innovations in products and processes as such. However, if British industry fails to adapt to new technology at least as fast as its overseas competitors, it will lose competitiveness and demand for its products will fall, with serious effects on employment levels.'

The group spells out its view on the 'fading away' in Britain of the manufacturing sector: 'The absolute decline of employment in the secondary sector in recent years was not inevitable nor need it be irreversible . . . The growth of employment in services is not necessarily associated in the long run with a loss of jobs in the secondary sector. The concept of de-industrialisation relates to the abandonment of given industrial activities as a result of a failure to compete effectively. It is the very opposite of the concept of a post-industrial society. Both may entail loss of industrial jobs; but in the former case it arises from shrinkage of the industrial base, and in the latter from a separation of the size of the industrial base (in terms of output) from the amount of employment it generates. The debate about microelectronics and employment has tended to obscure these two concepts.'

These two groups, then, see the advent of chip technology as a challenge to be seized by British firms. It could make them competitive once more. If the challenge is ignored, there could be mass unemployment, but the culprit will be de-industrialisation, not technology.

Other commentators see the situation differently. With the infusion of chip technology that is bound to arrive sooner or later, into most sectors of industry will come an unavoidable loss of jobs. Ian Barron and Ray Curnow in *The Future with Microelectronics* say that the technology will make redundant huge masses of workers involved in clerical and assembly operations. They are gloomy about the effects on employment of new products: 'It is hard to envisage new products, whether consumer or capital goods, which are of sufficient volume and value to absorb the labour displacement.' Barron and Curnow made this long-term assessment: 'Should the labour-displacement consequences (we have) put forward take place in a society experiencing low growth and balance of payments problems, we may be contemplating levels of unemployment at about ten to fifteen per cent of the workforce and more.'

Whether people are optimistic or pessimistic about the effects on employment of the new technology, they are generally agreed about one thing. Whatever happens, unemployment is unlikely to decrease significantly from today's levels that would be thought unacceptably high even a few years ago. And it could get worse – a lot worse. If anyone has got spare time on his hands in the microchip age, he could do a lot worse than to help plan for a society in which unemployment is a way of life.

Chapter 12: What sort of life?

If we *are* moving into a world where an overall slow-down in economic activity is accompanied by increased productivity due to computerisation, what will it mean for the individual? We have already seen that there will be a stronger likelihood of his being without work. However, the extent to which your job will be threatened, or your life as a whole change, will depend very much upon who you are.

Let us first consider more closely what we mean by unemployment. Even in Britain in the early 1980s, where more than two million are out of work, there are still many jobs left vacant because people are not there to fill them. Among the people in short supply are computer programmers, plumbers and decorators – groups of workers who could command high wages because of their scarcity. Yet on the other hand Britain has too many steelworkers, dockers, shipbuilders and low-skilled factory workers. All of these people have found in recent years that their jobs are fast disappearing. Anyone who investigates how to tackle unemployment must clearly do something about this mismatch between demand and supply in the various employment groups.

We have seen already how the proportion of technically-qualified people in the labour force is becoming higher all the time. That is, in a general sense, workers are becoming higher skilled. Yet this does not mean that skilled shop-floor workers will be immune from the unemployment effects of microelectronics. Professor Hiroyuki Yoshikawa of the University of Tokyo has compiled a revealing graph which shows the ease with which various kinds of workers can be automated out of a job by advances in computerised factories. The graph (see Figure 10) shows that workers concerned with assembly and machining operations, often relatively highly-skilled members

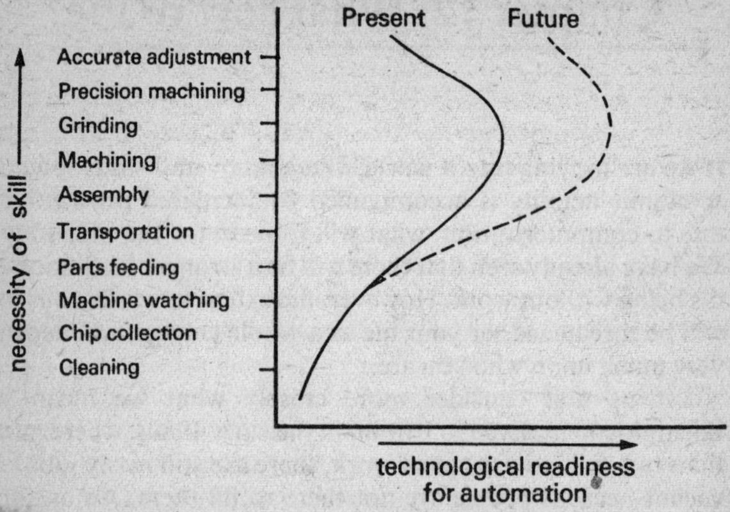

Fig. 10: Ease with which grades of labour can be automated out of existence.

Source: The Factory of the Future, Infotech

of a factory's workforce, can be replaced by machines much more easily than, say, cleaning staff. Only when a worker travels higher up the skill league table (towards the top of the graph) does the possibility that a computerised technique will take away his job start to diminish.

The veracity of the professor's graph, what we can call the skill-automation curve, is backed up by experience in Europe. According to the European Trade Union Institute: 'One tendency is for the introduction of microelectronics to significantly reduce employment in certain activities such as maintenance and repair work, assembly work and tool setting. This leads to a reduction in skilled manual employment. Within an enterprise this can be seen as a polarisation in employment between semi-skilled operatives on the one hand and highly-skilled technical staff on the other.'

To demonstrate how this happens, let us consider a specific firm, Standard Electric Lorenz, a European subsidiary of ITT. Among its products is an electronic telex machine in which one microprocessor has replaced no fewer than 936 parts in the device's electromechanical predecessor. As part of the move towards microprocessor-aided products, the proportion of factory-floor jobs in the company that require training fell from eighty-two per cent to thirty-five per cent. The proportion of semi-skilled jobs rose from fifteen per cent to thirty-five per cent and the proportion of highly-qualified technical jobs increased from just two per cent to thirty per cent of all employees.

This process of polarisation between semi-skilled and very highly-skilled people in the labour force is what Japan's Yoshikawa calls 'spotty automation'. According to his thesis, automation so far has taken away from people many of the reasonably interesting jobs in factories, leaving them the dirty, menial ones. It is interesting to look at Yoshikawa's plans to improve matters. He says that firms should push on with research into the totally unmanned factory (of which we mentioned something in Chapter 7) to make sure that machines are given the dirty jobs too.

Whether this approach will work depends on the rate of progress of engineers with robot systems that, for instance, recognise objects and feed them into machine tools. Yoshikawa, who is an adviser to Japan's national project on complex manufacturing systems, appears confident that the unmanned factory will soon be technically possible. And he has no doubts that it will be socially desirable. Unlike observers in Europe and the US, who are often worried about the extent to which such developments would deprive people of jobs, the Professor says that unmanned factories would suit the bulk of the population. His view is that as people become better educated and more highly-skilled, they simply will not put up with the kind of soul-destroying jobs in factories that earn so many people a living today.

If automation affects the skilled manual worker more than most, who else does it hit? In the office, it will affect female workers, who, as we have remarked already, have started to appear in the workforce in increasing numbers since the Second World War. The trend has not been confined to Britain. In Italy the number of women in the workforce climbed from twenty-seven per cent to thirty-one per cent over the ten years to 1977; in Sweden the growth was from thirty-seven per cent to forty-three per cent; in West Germany from thirty-seven per cent to thirty-eight per cent; while the figures for the UK are thirty-one per cent and thirty-nine per cent. Women are represented more strongly in the tertiary (service) sector than anywhere else. For example in the EEC as a whole, forty-five per cent of workers in banking and insurance are female; the figure for general commerce is forty-eight per cent. In any shake-out of employment in offices due to the introduction of computerised equipment, women would suffer disproportionately.

Other people hit particularly hard by the disappearance of job opportunities in industry are young people who have just left school. Lack of apprentice schemes, due to the melting away of craftsman-status and skilled-worker jobs in manufacturing industry, is partly to blame for this. The shadow of the jobless young has been cast over all of Europe. In the EEC nations, unemployed people aged under twenty-five who rose, as a percentage of the total numbers out of work, from twenty-eight per cent in 1973 to thirty-six per cent in 1978.

People who try to enter the workforce from school are naturally the ones to suffer in an overall squeeze on employment. And there are few demographic factors likely to diminish the size of the labour force as a whole. Certainly, by 1990 there should be fewer young people as a result of the effects of the early 1960s 'baby bust' when many in the developed world started giving up the idea of having babies. In the US, the total of fifteen- to twenty-four-year-olds should drop from forty-two million in 1980 to thirty-four million in

1990. But it will take a long time for the effects of the 'baby bust' to be felt throughout the whole population: in the US the numbers of people aged between twenty-five and sixty-five is expected to increase to the year 1990 and then flatten out, not falling until at least half-way through the 21st century.

It makes sense, it would seem, to teach people from an early age to get ready for the microchip era. They should be prepared at school for the kind of changes in employment patterns that we have just outlined. The UK's Department of Employment 1979 report on the implications of microelectronics makes just this point:

'The jobs that will be "wasted" will tend to be relatively low in skill content and this will mean that the brunt (of unemployment), if any, will largely be borne by school leavers or would-be re-entrants to the labour market looking for jobs with a relatively low skill content. If this turns out to be the case, the answer to the problem must lie mainly in the fields of initial education and training.'

The report goes on: 'In the long run information technology is likely to produce a new pattern of jobs and career opportunities which will require very radical rethinking of the school curriculum . . . New technology is likely to have the effect of reducing job opportunities for school leavers who lack basic skills of numeracy and literacy.'

Amazingly, the group that drew up the report was unable to find out from the Department of Education and Science exactly what was being done in schools to help to direct children towards a changing society based on chip processes and products. Britain's school system is so fragmented, with local authorities exerting almost total control over the schools in their area, that no one in the government department with responsibility for education could give the group's members a clear idea of what is happening. The dithering continued. It took over a year for the Conservative government, elected in May 1979, to ratify a plan published by the previous Labour administration to inject £12 million into schools in a special programme to boost awareness of microelectronics. The lack of urgency on the part of the British government contrasts with the situation in France, where in 1978 the government launched a programme to install at least one microcomputer in every school in the country.

Interestingly, the Department of Employment report ruled out the idea of big government re-training programmes to help displaced workers to find new employment. The study team said that firms to whom it spoke much preferred to do the job of retraining themselves. According to this view, broad-based government schemes were often a waste of effort because they do not give people the skills that companies require.

However, a counter argument against this is that in Britain, many computer firms who need specialist software writers fight shy of hiring people who need training. The companies feel that if skills are in such high demand, the trained worker will soon move to another job where a rival firm will receive the benefit of the original company's work in training him. The high cost of training to meet a changing world is borne out in a study by the Massachusetts Institute of Technology for the UK's Department of Industry on the application of microprocessors. 'Retraining . . . can consume time and money. Users face similar training costs. As the sophistication of the devices increases, this training aspect will pose greater challenges.'

Somewhere here we ought to offer resistance to the idea that job losses are always bad. In the view of American management theorist Peter Drucker, people ought to welcome redundancy, not fight it. He advocates the idea of governments 'planning for redundancy', the perfect way, so he says, of nations smoothing out the changeovers from one kind of employment pattern to another. Governments would actively encourage fading firms to close down, helping to find their workers jobs elsewhere. This would be instead of today's policy where governments often do all they can to prop up unsuccessful firms who employ a lot of people. Of course such a policy would not work in a vacuum. Both the government and the individual companies would have to commit themselves to the idea for it to work.

Effective planning for redundancy requires one of two alternatives. The first would be an assurance given to the workforce that, whatever happens, they will not be out of a job. People may be made redundant but they will not have to fear that there won't be another kind of job waiting for them. In this way, for instance, the transition could be made between service work (say, truck driving or canteen cleaning) in an unprofitable steel-making company and the same kind of activity in a booming consumer electronics firm.

This kind of policy is in force in Japan where many of the big companies have 'no redundancy' agreements with their workers. According to many observers, the policy breeds a sense of security amongst the workforce that makes them agree to technical changes and helps their companies commercially.

A consequence of this policy is that a certain amount of artificial employment will be generated. For instance, workers will be paid to mend fences or paint walls that are already in reasonably good repair – just to keep them off the unemployment register. However, such opportunities for artificial or non-commercially inspired employment could very useful. In Britain today, for example, there are huge numbers of tasks which need doing. They range from anything between digging old people's gardens to insulating houses. But in the absence of enough motivation on the part of companies to carry out these jobs – in other words there are no profits to be made – they are not done. A way of using the profits earned by more commercially orientated jobs to pay for socially useful activities would be welcome.

The second way of providing for people being made redundant is for society to effectively cease to recognise the difference between the states of being in work and being out of it. It means the abandonment of what social commentators call the 'Protestant work ethic'. In such a society, the profits earned by those who are in work would help support those unfortunate (or fortunate) enough to be without it. Thus there would be a choice between working or not. This statement may seem outlandish, because, some might say, surely no one would want to work if they could be paid for doing nothing. But is this really true? A lot of people work because they enjoy it, not just because employment gives them money for the necessities and pleasures of life. In a society where the rewards for working or not working were roughly the same, many would probably opt for a life of work, providing, that is, it is interesting and not soul-destroying drudgery. (This is where we return to the question of the social desirability of unmanned factories.)

Some people might argue that such a system is most extreme, and would destroy competitiveness, the term for the way in which people constantly move to and from jobs to earn more money. The 'rewards-for-not-working' society could encourage laziness and lead to economic stagnation, it could be argued. A compromise that would meet some of these objections would be a set-up that gave people long sabbaticals, a generally shorter working week and early retirement. In this way the amount of available work would be shared out among the population.

A feature of this society would be what could be described as 'negative income tax'. Today people pay income tax to the government depending on what they earn. The government then recycles this money to pay for projects that it deems socially desirable, for instance schools and hospitals, or for other projects whose very nature means that government has to direct them, for instance defence. Under negative income tax, some people would be taxed as they are now. But others whose revenue is low would have a negative tax – in other words they would be given money by the government. In this way everyone's income would be kept largely the same.

Of course, some might argue that this kind of arrangement is what we have already with hand-outs from government to the unemployed, the sick or the poor under the various benefit or rebate schemes. The important difference, however, is that under this new system both tax and rebates are handled as part of the same process. The government would be responsible for administering them as no more than a kind of clearing house through which the payments 'in' and 'out' are managed.

An important corollary to the scheme – indeed it would not work if this did not happen – would be for the people who *do* work to produce enough wealth to pay for the socially desirable 'unemployment-financing' schemes. This is where we return once more to the desirability of efficient industry. There would also have to exist a huge amount of social consensus so that the better-off do not mind financing payments made under the 'negative tax' schemes.

How could this happen? Commentators who want Britain to steer towards this type of society often conclude that the only possible way for it to work would be for all kinds of industry to take on board automated, microelectronics-aided techniques. With these, industry would be able to produce its goods efficiently enough to generate the revenue needed for the grand plan of social well-being. The inference is that if companies do not get to grips with these ideas and channel them into products and processes, they will not produce enough to keep even their most easily satisfied shareholder happy, let alone people in the rest of the country who might be depending on the firms' good performance for their next meal.

In the scheme just outlined the government acts as a kind of national banker to distribute wealth earned by successful industries among the population as a whole. Japan's way of running its industries produces an effect that is somewhat similar to this. However the method is very different. There is no State interference of any sort. Instead, people who work for successful manufacturing firms earn good wages, and are protected from unemployment in slumps by the 'no redundancy' agreements that their employers institute. The people then spend their (high) wages on the products of other industries that are not nearly as efficient – for example service industries – and which therefore have to charge high prices. But this does not matter as people can afford to pay them. The practice provides a good example of a practical way of ironing out the differences between a country's efficient and inefficient businesses so that the population benefits.

It is worth quoting an official UK report that delves into the operation of the Japanese service industries – and the way this differs from practice in Britain. The passage is from a January 1980 report from the Advisory Council for Applied Research and Development called *Technological Change: Threats and Opportunities for the United Kingdom*.

'It is clear that the future success of much of the UK manufacturing sector depends on improving productivity and reducing manning levels. By contrast, the UK service sector, despite a post-war growth of productivity lower than in manufacturing, uses labour relatively efficiently by international standards. In deployment of labour, the UK stands at the opposite extreme from Japan, where manufacturing industry has a positive commitment to reducing its labour force but where for example distribution, a major service industry, absorbs twenty-two per cent of the national labour force. This is almost double the percentage in the UK, and almost as much as the Japanese manufacturing sector.

'The Japanese appear to reap two important benefits from this aspect of their employment policies. First, because the majority of manufactures are traded internationally and the majority of services are not, over-manning in the service sector does very much less harm to the overseas competitiveness of the Japanese economy than over-manning in UK manufacturing does to our own. Secondly, there is much greater benefit in terms of the quality of the product from increased employment in service occupations than in production.

'Modern technology makes possible substantial reductions in the number of people involved in production and will usually improve the quality of the product. By contrast, almost every kind of service would be improved if there were two people working for every one currently employed – shorter queues at retail check-outs, more rapid attention to customers' queries by sales departments, more personal service in hotels and restaurants etc. Services would cost more – and they are relatively expensive in Japan – but manufactured goods would be cheaper both absolutely and by comparison with imported goods. Home and overseas sales of manufactures would expand and the dearer services could be afforded.'

In this scenario, then, super-efficient manufacturing industries earn the revenue to pay for improved services. People find work in these areas and so overall employment is not reduced. But – and this is a very real but – what happens to Britain if the manufacturing firms do not improve in performance to the degree required? Instead of this, they get even worse, gradually becoming less competitive, shedding workers and closing down. There is then no 'safety net' of wealth produced by successful industries to pay for the desirable and, by their very nature, inefficient service industries that people hanker after. The difference between Japan and Britain is that the former has this safety net; the latter is struggling to obtain one, although all the signs, from looking at previous attempts to raise efficiency in manufacturing, are that it will remain an illusion.

In a country where manufacturing continues to decline, throwing large numbers of people out of work, the government could adopt a very positive policy. Instead of sitting back and doing nothing, it realises that the idea of breathing life into manufacturing is hopeless, so it would start to re-educate people to accept lives without jobs.

The new burden on the elected administration, once it had got the message that it could no longer manage the responsibility, handed over by Keynes, of maintaining full employment, would be to maintain happiness within unemployment. Education for leisure would be the new order of the day.

This new burden would involve mass programmes in sport, physical education or do-it-yourself. People would be encouraged to indulge in educational activities of all kinds. It would be a step towards what some people call the 'knowledge society'. The new order would be the only way out of a huge economic and social mess. The government would do all it could to channel workers from the old industries into the new ones, based on the spread of education and electronic transfer and collation of information.

To those of you who find such a prospect alarming or smacking of a dictatorship, relax: it is never likely to happen. Governments are much too reactionary to plan for, let alone implement such sweeping changes, even though these changes would probably be most beneficial.

Instead, in Britain we are more likely to see the appearance of a rather different society. This is a society based absolutely on the increased polarisation between the technological élite and the relatively skill-less workforce. This parallels what is already happening in factories and in office work. It also echoes the ideas sketched out in Chapter 6, of the two halves of the population – computer 'haves' and 'have-nots'. Just as the computer 'haves' – the people who understand the devices – have the best of what social life there is in the introverted electronic society that might be round the corner, so they would have the best of it at work.

The technically élite, the only ones who know how to control the forces that muscle in on everyone's life, would be in the upper echelons of management. Their offices would be command posts from which they survey the only partially, or 'spottily', automated factory-floors of what is left of manufacturing industry. As has so far happened, the machines would do all the interesting jobs, the humans the boring ones. The workforce would be an uninterested, ill-satisfied corps of machine minders.

For all the technical sophistication apparent in the factories, in the offices that mushroom everywhere to support clerical service employment, and in people's homes, which fairly bristle with mindless and useless consumer gimmickry, the society is not a healthy one. It is at odds with itself because of the extreme inequalities that it engenders. The split society can't paper over the cracks. It gradually breeds more discontent and in due course the breaking point is reached as the society falls apart in an explosion of hatred. This is what, conceivably, could happen.

On the way to such a break-up, the pervasiveness of government and monopoly capital would increase. Government's influence would insidiously become larger, helped by the use of computers to keep files on every aspect of the population over which it rules. Computerised communications would be one of the features of this future. Data networks would everywhere flash information from one point to another, controlled not by the population as a whole but by the élite who run the microchip-based society.

Fears about privacy intrusion as a result of this kind of development have long been voiced in the developed world. Many Western countries, among them Sweden, the US, France, Norway, Belgium, Austria and Luxembourg, have laws to regulate what can and cannot be stored in computerised data bases. The computers could be used for undesirable ends. One such nasty trait would be for governments or employers to keep files on politically-orientated people to stop them getting jobs. Or governments could pass computerised information, stored for a quite legitimate reason, for instance on an individual's tax records, to the police to try to effect a prosecution.

Britain has been talking about instituting a data protection law for an incredibly long time. In February 1970 the Home Secretary, James Callaghan, set up a committe under Kenneth Younger to examine the whole question of privacy. In its report the committee highlighted the growing use of computers. Eventually in December 1975, the government published a White Paper based on Younger's views which stated:

'The time has come when those who use computers to handle personal information, however responsible they are, can no longer remain the sole judges of whether their own systems adequately safeguard privacy.'

As a result of the White Paper, the government appointed yet another official body, the data protection committee, to investigate how the desired safeguards could be put into practice. The committee, chaired by Sir Norman Lindop, eventually reported in December 1978. Its 460-page report emphasised the need for a body, to be called the Data Protection Authority, which would regulate the kinds of information put into computers. All the signs in 1980 were that the committee's report had joined the others on the Whitehall bookshelves marked for studies which the government has no intention of acting upon. In September 1979 Lindop gave vent to his dissatisfaction that nothing had happened (the government officially said it was 'consulting' people whom the proposals would affect). In a speech in Edinburgh, he said the implications of the spread of computer data systems were 'too serious to be left to chance'. People's personal details sorted in computerised files could lead to some 'pretty frightening developments' that would happen 'quite quickly without most people being aware of what was happening'.

A worrying fact about many computerised data bases is the frequency with which they contain errors. Thus a society based on the use of such instruments would often base decisions on completely wrong information. A study by Kenneth Laudon of the City University of New York, presented to the American Association for the Advancement of Science's 1979 meeting in Houston, gave an alarming account of error-strewn data bases. People on social security in the US have a thirteen per cent chance of the computer that details their records logging an inaccurate piece of information. If they happen to live in Boston or Denver, the error rate is as high as seventeen per cent, according to Laudon's figures.

Laudon also looked at criminal records. This is an important area in view of the fact that data on known criminals, stored in one police force's computers, can easily be sent by telecommunications links to forces in other parts of the country, or even to police departments abroad. In the state of New York, Laudon reported that only twenty-seven per cent of criminal records were completely accurate. Many of the records were ambiguous or incomplete. Laudon said the findings were 'outlandish' in the light of the degree to which the US plans to use its computerised criminal history system. According to Laudon, by 1985 almost half of the criminal justice decisions taken in the US will be assisted by the system.

Side by side with the move towards state interference, via computers, with many aspects of life, would come the growing dominance of big computer firms. Already a small number of companies exert a big say over what happens in today's computer world. IBM, for instance, is responsible for some sixty per cent of the world's installed big computers. Intel exerts a similar, though not quite as all-enveloping, influence on the microprocessor world. Once a computer user has purchased a certain manufacturer's machine, he often finds it difficult to break away. This is because the user gears the whole of his operation – that means software and other services – to the manufacturer's type of machine. If he wants to buy another firm's product he may often have to discard years of effort building up software that will be useless on the new machine.

Firms which have this near-monopoly will do all they can to perpetuate it, making sure that their machines will not run on standard software that applies to other manufacturers' computers. Users of international networks, for example, have tried for years to press IBM to become enthusiastic about a software plan that would mean computers of any make, anywhere in the world, could communicate directly with each other. But IBM has always been reluctant to get involved in this activity which could weaken it commercially.

Of course, it's possible to flip the coin and turn over some healthy developments that arise from a computer-orientated society. Increased use of computer communications could reduce the need for travel, so reducing energy costs in a world that now has to put aside a big proportion of its spending on energy. Some people have postulated that computer conferencing, via data and video telecommunications links, could result in many advantages. People might be able to work at home instead of travelling long distances to their place of employment. Executives could confer remotely via these links with a foreign business contact. This would save people having to jet across the world, just to sit the other side of an office table with a business associate. Developments in communications could benefit elderly, or handicapped, people who would feel less cut off from the outside world.

For all the optimism displayed by some people that computers can cut travelling costs, history is not on their side. Advances in telecommunications, albeit not computer-assisted, have been one of the most significant developments of this century. But they do not seem to have curbed people's demand for travelling. As each new telephone is installed the number of people queuing for jet flights grows, and airlines' demand for scarce space for new runways gets bigger too.

So far this book has concentrated almost exclusively on the developed countries. It has looked at the people who already have a toehold in the world of computers. What of the vast tracts of the earth inhabited by people who have hardly heard of a telephone, let alone a microprocessor? It is not for me to discuss here in depth the Third World's future in the computerised era. Yet all the signs are that it doesn't look too healthy. As the industrialised countries gradually become converted to more efficient, chip-aided industries, the gap between the rich and the poor of the world will probably get bigger.

Rising energy costs will almost certainly further this trend. If the wealthy nations of Western Europe are getting worried about how to pay for their oil imports, the situation must be even worse for a struggling country trying to pull itself into the twentieth century with energy-intensive methods. And as computers start to play a part in developing nations, the danger is that they will do so along guidelines established by the big Western electronics companies. IBM, for instance, has got a toehold in many emerging nations which depend almost entirely on it for what computer expertise they do have.

Over the past few chapters, we have sketched in some of the characteristics that will be part of life over the next fifty years. Some may be good, some bad. But what has been described will not happen automatically. Much depends on the policies that nations adopt – not in ten years, but now. The closing chapter will look at four options that we have as we enter the microchip age.

Chapter 13: The future

There are several choices for a country like Britain as it looks into the future. They may be similar for some of the other countries in the developed world. Britain's position is made rather different, however, by two things. The first is the extent of the de-industrialisation that has affected the country in recent years and has led to an extremely marked slow-down in manufacturing activity. The second is the presence of North Sea oil. The oil could give a tremendous boost to the nation during the 1980s. If the correct decisions are made, the reserves could help Britain to plan properly for the future.

During the 1980s the country will be self-sufficient in oil. By 1985, government revenues due to the exploitation of the commodity could amount to as much as £20 billion annually, boosting gross domestic product by some twelve per cent at 1980 prices. The extent of the advantage that this gives Britain may be assessed by considering the likely oil import bill for other industrialised countries over the next few years. For instance, in 1980 the US expects to spend £31 billion on imported oil, Japan £20 billion, West Germany £10 billion and France £8 million. Back in 1973, before the UK was thought of as an oil-rich country, it imported 113 million tonnes of oil: now it needs to buy virtually nothing.

The benefits from the oil could all too easily be frittered away, however. The extra cash that it brings to the country, instead of being invested in socially useful programmes, or in the industries that are going to be important over the next fifty years, could be wasted. Even worse, the cash might never be noticed at all if Britain's industrial output continues to stagnate. The money might purely compensate for deficiencies in other areas. When the crutch of the oil reserves is pulled away, as will happen when the oilfields start to dry up during the 1990s (providing no new reserves are found in the meantime), Britain's position could be harshly exposed – as that of a country in continuous and, by then, probably absolute, decline.

Let us look at the four broad directions in which the country can steer. These scenarios of future society could be expected to apply to Britain around the year 1990. They are listed in roughly the ascending order of radicalism that would be needed to bring them about.

1. The patched-up go-for-growth society

This is very much a continuation of what we have already. The country introduces, at a leisurely rate, devices such as robots and other automated equipment into what is left of its manufacturing industry. The industry becomes increasingly uncompetitive, however, compared with other nations. 'Spotty automation', a term introduced in the last chapter, becomes the order of the day.

Reasonably successful manufacturing firms, such as those concerned with electronics, continue to progress in terms of making reasonable profits. But even the 'go-ahead' electronics companies do not adapt quickly enough to the requirements of consumers who, for several decades, have been used to the idea that their standard of living goes up year by year. Thus people want more and more goods – for instance in the area of computerised gadgets for the home – of the kind that can only be purchased from foreign firms. The same is true of other kinds of manufacturing industry. Eventually companies in this area become so uncompetitive that more and more close down. Britain's industrial base shrinks to almost nothing apart from a small, select number of imaginative, innovative companies.

The service sector of the economy continues to expand, although at a slower rate as governments squeeze the services that they pay for – for instance in teaching, local council work, social services and hospitals. The nation gets into increased balance of payments difficulties as a result of the massive import bill. There is increased international unease about Britain's position in the world. Meanwhile in Britain the society rapidly splits into two increasingly-polarised camps, the computer 'haves' and the computer 'have-nots', as the manifestations of the microchip age make their presence felt.

Economic outlook: exceedingly gloomy. Overall state of happiness: fair. Prosperity level: very reasonable for some, low for the great majority (with prospects of getting lower). Likelihood of all this taking place: high.

2. The robot society

All levels of industry really wake up to the usefulness of microelectronics, much as they have been exhorted to by the UK government in 1978 and 1979. In a rush of enthusiasm and energy, companies invest heavily in automated equipment. Amazingly, they find the cash to do this from orthodox sources of finance, for example the London Stock Exchange, which abandons past practice and realises that financing technology is the key to a sound future.

Trade unions are won over to this approach, mainly by the use of what are called 'technology agreements'. With great rapidity, that owes nothing to the drawn-out way in which management–union negotiations are normally carried out, firms conclude these agreements with their work forces. They stipulate that new technology shall only be introduced with the consent of workers. Everyone knows this means a certain amount of lay-offs; but the trade union movement agrees to the policy in the interests of the long term. There is some suffering on the part of displaced workers and their families. But the effect of this is lessened by an enlightened government attitude which raises unemployment pay and institutes a certain amount of re-education, planning for leisure and so on.

In the lead-up to full automation, selective and temporary

import controls are imposed to let industry get on its feet. These are imposed mainly to placate the TUC, and they lead to a reduction in the volume of unemployment that might otherwise be expected. They are meant to be relaxed as soon as industry is felt to be generally strong.

In the service sector, steady growth occurs in private industry. Much of this is helped by computerised equipment which results in more efficient banks, insurance firms, shops and offices. However, it also results in some loss of jobs. In the public services, it is a different story. There is a heavy reduction in the amount of work they do as the government warns everyone to 'tighten belts'. Thus many of the services people are used to either disappear or are cut down. People have to pay if they are unlucky enough to go to hospital more than twice in a year; the refuse men call only once a fortnight; ambulance men answer 999 requests only if the victim has a computerised bank account with them; and people whose house goes up in smoke, necessitating the calling out of more than one fire appliance, have to pay a surcharge.

The nation puts up with this approach with the official exhortation: 'Grin and bear it: it'll work out all right in the end.' The vision is that eventually Britain will have a successful service-orientated economy along the lines of Japan's. The belt is loosened only in the late 1990s by which time the nation is in a healthier state economically. By this time a series of fifty or sixty nuclear power stations, which the government ordered during the 1980s, are churning out a good part of the country's energy requirements as North Sea oil begins to wane. There is a constant fear, however, that they will go wrong. This adds considerably to the number of people in towns near to nuclear power stations who seek psychiatric treatment.

Unemployment throughout the 1990s is high, amounting to ten per cent or so of the working population. Life for the three or four million workless people and their families is a bit grim. But then, as everyone else says, they are sacrifices in the cause of progress.

Economic outlook: good. Overall state of happiness: patchy. Prosperity level: reasonable. Likelihood: quite possible.

3. The 'Britain at war' society

The country is not at war with anyone. What it is fighting are the deeply ingrained attitudes engendered by the past. The 'new' society in Britain asks people to forget any ideas they had about the greatness of their industrial heritage. Consequently, the country steadily abandons many of the manufacturing industries that made it famous in the 19th century. Britain is now ready to face up to a new order of things: if other countries can make goods better than us, why bother to compete? That, not competition at all costs, is the new maxim. Therefore importing virtually all manufactured goods is an accepted way of life.

The new society also asks its members to forget all they once believed about the thesis that everyone must be 'gainfully employed'. This, it says, is nonsense. People should work only if they want to: others should be encouraged to pursue activities that they find satisfying, whether an organisation pays them or not. Everyone is guaranteed a living income by a government-run 'negative tax' system of the sort outlined in Chapter 12. This would have the effect of greatly increasing unemployment pay. There would be no backtracking on public services, even if the country was in a bit of a mess economically. In fact health expenditure, social aid schemes, employment programmes and so on would all be increased as part of the effort to help people to adjust to a new way of life. The government would spend a large part of its budget on 'educating for leisure', helping people to plan what they will do when they retire, for instance. Defence spending would be cut to pay for this. Clubs for all kinds of activity would be supported far more enthusiastically, by a wider membership, than is now the case. It would be a 'doing' rather than a 'working' society.

Service industries, especially those that involve computers and other automated techniques, would be encouraged. People would be exhorted by government to set up businesses, or get jobs, in what are called the 'knowledge industries'. In these, people write programs for computers in almost every conceivable area of life, prepare teaching material, write books, journals and pamphlets, prepare electronic information schemes, or work on imaginative ideas such as how to tap the seas or space for new sources of minerals or energy.

Gradually the country would find export markets for much of these knowledge-based goods. In particular, developing nations would be much seduced by the idea that these products provide the best ways of improving their economic status.

Britain would, inevitably, run into balance of payments difficulties. All too easily the import bill in manufactured goods could become enormous. However the problems would be kept in check by high taxation which helps to pay for the social services. Revenue from tax as a result of North Sea oil helps as well.

In general – and this is another change from today's society – people do not get too worried about the degree to which the country's failing to make ends meet (from an international viewpoint) finances the change to a new way of living. There is enough confidence that eventually the nation will be healthy enough to repay the money it has borrowed abroad. Just as in a real war, when governments think nothing of running up foreign debts to pay for armaments, the cash is needed for some big battles.

Economic outlook: good. Overall state of happiness: high. Prosperity: reasonable. Likelihood: low.

4. Low-growth, high-ecology society

People are determined to cut their dependence on energy and to use technology that is related to human skills. The overruling thesis is to increase happiness and personal satisfaction. So chip-aided techniques are frowned upon if they are applied in mass-manufacturing operations and the like. However, personal computers, chip gadgets for the handicapped and other devices that help the individual are all welcome.

The society institutes permanent import controls on many kinds of manufactured goods. Of course this means foreign countries place their own import controls in retaliation, but this does not matter all that much because there is very little that Britain makes in which anyone else is interested. People are so busy thinking about how to increase personal satisfaction that they don't have time for dreaming about the luxuries they are missing out on. The country is poor, but happy. Consumer goods are a luxury. For instance, there is a grave shortage of things that at one time people used to rely on to be imported from abroad. This applies to products such as cars and TV sets, the last UK vestiges of these industries having finally given up the ghost a few years before the 'low growth' policy was instituted. Again, this does not matter because a little austerity does wonders for personal satisfaction.

In this kind of society, manufacturing industry is by no means finished. Of course, it's a little unfamiliar to people brought up with the notion of modern factories. All over the country there are people engaged in a new style of industry: they make bridles for horses, build houses out of planks of wood, make pots and pans and rudimentary bits of china from metal and clay, and even occasionally turn their hands to forging tools in primitive furnaces.

It is very much a society built on renewable resources. Citizens mend their own battered old petrol-powered turbines to provide trickling amounts of electricity. Maintenance workers are in great demand for repairing the steam engines in the ramshackle buildings that pass for workshops. People hardly throw anything away: for instance, old food cans come in handy for cooking containers and packaging is used for writing paper. Windmills and solar collectors constitute the main way of providing energy. Although society's energy needs are small (it is just as well as these rude bits of equipment do not provide very much) the windmills and collectors are everywhere in evidence. They clutter up the countryside to such an extent that some people complain that the structures interfere with the country's main social activity, mass jogging (it helps everyone to keep warm), and also with the important pastime of hunting hares and foxes, animals on which people are forced to depend for food, meat being in short supply as a result of the import clamps.

Economic outlook: poor. Overall state of happiness: high. Prosperity: low. Likelihood: low.

Exactly what shape the future takes depends very much on what happens over the next few years. If you don't like the look of one of these scenarios but think it might be coming, you had better start doing something about it. What, however, of the future for the tiny chips that are so much bound up with our view of how events will turn out? Can the circuit elements that they contain continue their 'shrinking act' indefinitely?

A pointer to the answer to this question came in 1979. Engineers and scientists from IBM described how they had managed to define lines on a silicon substrate that were just eight nanometres wide. This is a distance of eight billionths of a metre, or about fifty times the diameter of an atom. No one had succeeded in building such tiny circuits before. (By contrast the definition in 1979-vintage semiconductor circuits are, at the smallest, about 2000 nanometres.) If scientists can continue on the path of smaller and smaller circuits, it will not be long before chips that contain one million, then one billion, circuit elements appear on the scene. That such things will appear one day is scarcely open to question. It is all a question of 'when'. What people do with such 'super chips' is, however, much more open to conjecture. They could use the devices to work out how to send mankind to the outer planets of the Solar System, or how to conquer disease. Just as easily, people could harness such chips, which, by now, would be as intelligent as several

Einsteins rolled into a piece of material the size of a wheat ear, to wage war or bring about tremendous social discord.

Whatever happens, no matter how clever or intelligent collections of microcircuits become, the overriding requirement is for people not to trust them too much. This may seem strange, after all these chapters that say how wonderful cheap computers are. However, computers do go wrong. They sometimes make decisions incorrectly. To leave important areas of the running of the world to computers would be a disaster. Of all the possibilities as we enter the age of cheap computers, the most distasteful is that people are so blinded by the power of the machines that they forget to supervise them.

References

The following is a list of books and reports consulted, broken down into three headings, (a) general, (b) applications and (c) social economic effects.

A. GENERAL

The Microprocessor Industry, Creative Strategies International, 1979

Microelectronics, reprints from *Scientific American*, W. H. Allen, 1977

Chips in the 1980s, Economist Intelligence Unit, 1979

Revolution in Miniature by Ernest Braun and Stuart MacDonald, Cambridge University Press, 1978

Introducing Microprocessors by G. L. Simons, National Computing Centre, 1979

The Future with Microelectronics, by Iann Barron and Ray Curnow, Frances Pinter, 1979

Microelectronics Into the 80s, Mackintosh Publications, 1979

The Electronics Industry in Scotland, Booz Allen and Hamilton, 1979

Tomorrow in World Electronics, proceedings of Financial Times Conference, March 79

The Microelectronics Industry, NEDO, 1978

The Applications of Semiconductor Technology, report by Advisory Council for Applied Research and Development, HMSO, 1978

B. APPLICATIONS

History of Manchester Computers, by Simon Lavington, National Computing Centre, 1975

Electronic Computers by S. Hollingdale and G. Toothill, Penguin, 1976

Automation, report by Department of Scientific and Industrial Research, HMSO, 1956

Joining and Assembly, report by Advisory Council for Applied Research and Development, HMSO, 1979

National Development Programme in Computer Assisted Learning, Final Report of the Director, Council for Educational Technology, 1977

Report of the National Committee on Computer Networks, HMSO, 1978

Telecommunications in the 1980s and after, Transactions of the Royal Society, London, 1978

The Modernisation of Telecommunications, Post Office Engineering Union, 1979

Data Communications Outlook in Europe, PA Management Consultants and Quantum Science Corporation, 1974
Telecommunications for Metropolitan Areas, US National Academy of Sciences, 1978

C. ECONOMIC/SOCIAL EFFECTS

TUC Economic Review 1979
Impact of Microelectronics on Employment in Western Europe, European Trade Union Institute, 1979
Manpower Implications of Microelectronic Technology, report by Department of Employment, HMSO, 1979
Microprocessor Applications: Cases and Observations, report prepared for Department of Industry by MIT, 1979
Report of the Committee on Data Protection, HMSO, 1978
Industrial Innovation, report by the Advisory Council for Applied Research and Development (ACARD), HMSO, 1978
Technological Change: Threats and Opportunities for the UK, report by ACARD, HMSO, 1980
Technology and the Employment Problem by Ian Mackintosh, contribution to Financial Times Conference, November 1979
Technical Change and Employment by Ray Rothwell and Walter Zegfeld, Frances Pinter, 1979
The Collapse of Work by Clive Jenkins and Barrie Sherman, Eyre Methuen, 1979
The UK Economy: A Manual of Applied Economics by A. Prest and D. Coppock, Weidenfeld and Nicolson, 1978
Office Technology: the Trade Union Response, Association of Professional, Executive, Clerical and Computer Staff, 1979
Social Implications of Microelectronics, Central Policy Review Staff, 1978

Index

GOOD WORK

BY E. F. SCHUMACHER

WHAT IS THE PURPOSE OF OUR WORK?

When Dr Schumacher took a look at the plight of businesses today and said 'small is beautiful' he spoke to millions and coined a phrase. In GOOD WORK he addresses a question which is central to most of us and one which is all too often ignored by the economic structure of the Western world. Dr Schumacher maintains that the purpose of man's work is threefold: to produce necessary, useful goods and services; to enable us to use and perfect our gifts and skills and, finally, to serve and collaborate with other people in order to liberate ourselves from inbuilt egocentricity. A job in which one finds no personal satisfaction destroys the soul. With sanity and sensitivity the late E. F. Schumacher offers important and thought-provoking alternatives which point the way to mankind's physical and mental liberation.

'Compulsive reading'
TIME OUT

'The message of this set of essays is beguiling; technology is not the inevitable ruler and real alternatives seem possible'
NEW SCIENTIST

ECONOMICS 0 349 13133 3 £1.95

ALIEN INTELLIGENCE

BY STUART HOLROYD
(Illustrated)

IS THERE ALIEN LIFE ON EARTH?

Bestselling author Stuart Holroyd begins his examination of alternative intelligence with some of its earthly manifestations – the whales, dolphins and chimpanzees who are all capable of abstract thought.

He analyses ESP, ectoplasmic phenomena, out-of-the-body experiences, and the possibilities of alien mind-control, and comes to the startling conclusion that the mind has an independent and immortal psychic existence.

He investigates the claims to intelligence of artificial minds and thinking machines, together with the archetypes of myth and pre-history that hold the clue to alien visitations. These staggering revelations invite the reader to share the author's compelling belief that there is alien life on earth.

COSMOLOGY 0 349 11709 8 £2.50